Whales and Elephants in International Conservation Law and Politics

Whales and elephants are iconic giants of the marine and terrestrial animal world. Both are conspicuous representatives of wildlife conservation. The issues of whaling and the ivory trade are closely linked, both legally and politically, in many ways; some obvious and some surprising. The treatment of both whales and elephants will be politically and legally contentious for years to come, and is of great significance to conservation in general.

This book examines the current state of international environmental law and wildlife conservation through a comparative analysis of the treatment of whales and elephants. In particular, it describes the separate histories of international governance of both whales and elephants, presenting the various treaties through which conservation has been implemented. It is shown that international environmental law is influenced and shaped by important political actors – many with opposing views on how best conservation, and sustainable development, principles are to be implemented. Modern environmental treaties are changing as weaknesses and loopholes are exposed in older, and possibly outdated, treaties such as the Convention on International Trade in Endangered Species of Wild Fauna and Flora (CITES) and the International Convention for the Regulation of Whaling (ICRW). Such weaknesses can be seen in the efforts made by some states to circumvent or weaken CITES and the International Whaling Commission and to resume commercial whaling; and further in the efforts of countries to resume trade in ivory. The argument is made that the Convention on Biological Diversity could be used to begin reconciling opposed views and to focus conservation efforts.

The argument is made that effective conservation of species cannot be achieved through individual treaties, but only through a synergistic approach involving multilateral environmental agreements – 'ecosystems of legal instruments'.

Ed Couzens holds the degrees of BA Hons LLB (Wits), LLM Environmental Law (Natal and Nottingham) and PhD (KZN). He is an Attorney of the High Court, South Africa; and, at the time of writing, an Associate Professor in the School of Law, University of KwaZulu-Natal, Durban, South Africa. He is rated as an 'Established Researcher' by the National Research Foundation of South Africa for the years 2013–2018; has been an Assistant Editor on the *South African Journal of Environmental Law and Policy* since 2001; and has been a co-editor of the University of Eastern Finland/United Nations Environment Programme *International Environmental Law-making and Diplomacy Review* since 2006. He is also a qualified field guide in South Africa.

Whales and Elephants in International Conservation Law and Politics

A comparative study

Ed Couzens

Routledge
Taylor & Francis Group
LONDON AND NEW YORK

earthscan
from Routledge

First published 2014
by Routledge
2 Park Square, Milton Park, Abingdon, Oxon OX14 4RN

and by Routledge
711 Third Avenue, New York, NY 10017

Routledge is an imprint of the Taylor & Francis Group, an informa business

British Library Cataloguing-in-Publication Data
A catalogue record for this book is available from the British Library

Library of Congress Cataloging-in-Publication Data
Couzens, Ed, author.
Whales and elephants in international conservation law and politics : a comparative study / Ed Couzens.
pages cm
Includes bibliographical references and index.
1. Wildlife conservation (International law) 2. Whales–Conservation–Law and legislation. 3. Elephants–Conservation–Law and legislation. I. Title.
K3525.C68 2013
346.04'695416–dc23
2013020410

ISBN: 978-0-415-65905-5 (hbk)
ISBN: 978-0-203-07533-3 (ebk)

Typeset in Galliard
by Keystroke, Station Road, Codsall, Wolverhampton

MIX
Paper from responsible sources
FSC® C013056

Printed and bound in Great Britain by
TJ International Ltd, Padstow, Cornwall

Contents

Acknowledgements and dedications

Over the years there have been so many people in so many countries who have assisted me with my research, encouraged me and given of their time and knowledge, that it would be impossible for me to mention them all. All of the staff at the IWC Secretariat were wonderful, and made sure that the only research difficulty I had was in finding Impington and then the Red House itself. Different Ministers of Environmental Affairs kindly gave me permission to join South African delegations to the IWC. The people at Earthscan Routledge, particularly Tim Hardwick and Ashley Wright, have been extremely professional and at the same time very patient, as were Maggie Lindsey-Jones and other members of the Keystroke production team.

Friends and mentors abound. I have learned much about environmental law, international law and diplomacy from all of Herman Oosthuizen, who welcomed me when I cheekily asked to attend the IWC (but my gratitude does not extend to being convinced that we saw a sea otter off Kenai!); Tuomas Kuokkanen, who took an amazing gamble on me despite (or perhaps because of!) the vodka incident; Jeremy Ridl, the best environmental lawyer in practice (as well as many more dubious talents); and Michael Kidd, the best environmental lawyer in academia and who became the supervisor of the thesis which preceded this book (for his sins, but it's his fault for dragging me into academia in the first place!). Certain other friends, colleagues and partners-in-crime have also supported me beyond the call of duty over the years: Adrian Bellengère; Andrew Blackmore; Marita Carnelley; Cam Carruthers; Andrea Gabriel; Melissa Lewis; Devarasi Maduramuthu; Alan Northover; Li Steyn; Meaghan Sunderland. My famil(ies) in Sefrica, Sweden and Transylvania have always supported me – Katty Couzens, Kate Couzens-Bohlin and Hal Couzens especially. Meda-Mihaela Couzens supports me unswervingly whilst being my fiercest critic – I am very proud to be married to her. And finally, my hope is that one day Kay Couzens will look back and say that his tati's heart was in the right place . . . even though I knew so little about nature that I struggled to see the difference between cows and horses (let alone between balenas and fantoos).

Ed Couzens
March 2013

1 Introduction

> . . . all comparison in the way of general bulk between the whale and the elephant is preposterous . . .;
> nevertheless, there are not wanting some points of curious similitude . . .
>
> Herman Melville *Moby Dick*

Useful understanding of international conservation law and politics can be gained from considering two multilateral treaties – the International Convention for the Regulation of Whaling (ICRW) and the Convention on International Trade in Endangered Species of Wild Fauna and Flora (CITES); and between whales and elephants, high profile species which the two conventions regulate in international law. There are certain connexions between the conventions and between the species which are of greater importance than might appear from a superficial consideration, and which have a place in the ongoing debate over which philosophy of use of natural resources the world should adopt. Understanding how and why these connexions exist can help us better to understand the debate, the politics and the international law in the area.

The argument in this book is essentially that whales and elephants have, through the histories of mankind's interactions with them, reached similar points within international law – despite these species' inherently different statuses either as free for taking (whales) or as falling into the dominion of states (elephants). The protections afforded to both species are argued by many not to be adequate, or not to be sufficiently nuanced to be effective, with these arguments being made both by proponents of preservation (in the sense of near-complete protection) and proponents of conservation (in the sense of sustainable use). An important reason why the management, in international law, of both species is not currently optimal is that the separate treaties (the ICRW and CITES) which govern such management in international law are inadequate. The two treaties are 'elderly' and their respective operations are riven with conflict. There are essentially two reasons why the two treaties are currently inadequate. These are that both are representative of a problematic tradition of managing particular wild species on their own, as if these species were somehow isolated from their ecosystems, and that the treaties both suffer from the problem that their states parties currently lack the common vision necessary for the treaties to succeed. This becomes apparent from study of efforts made at meetings of both treaties to undermine (or at least to circumvent) the other, and of the stark differences in beliefs which are revealed at such meetings. Conventions are always products both of their histories and of their particular periods.

The conflict between the parties sometimes seems irresolvable, given the deep philosophical differences which appear to exist. It might eventually be possible, however, to reconcile the different, even seemingly mutually exclusive, views of states under the framework of a rather different sort of treaty: an all-encompassing treaty which considers all of the component parts of biological diversity – and it is suggested that the world is indeed moving, albeit slowly, in this direction. Whales and elephants have come, through their histories and perhaps because of their natures, to be focal points for the world's states in the battle to determine what mankind's future relationship with all wild species, and all natural resources, should be. The battle over whales and elephants therefore has significant implications for, and may even help to determine, the shape of this future relationship.

Both species (referring to elephants and whales generally rather than differentiating sub-species) are perceived by many to be endangered. Both probably play 'keystone' (or even 'super-keystone') roles within their environments. Both species are generally considered to be 'especially' intelligent and sensitive. Both species have experienced near-catastrophic population declines due to overuse by humans and both have come, over time, to be given relatively stringent protection in international law. There are then less obvious links, such as both species (or at least sub-species or populations thereof) being argued by different states to be non-endangered, with the concomitant argument being made that exploitation might be a positive thing for long-term survival. The point here is that the two species share the somewhat peculiar status of being perceived simultaneously as both endangered and non-endangered (even over-populous). Further, and very specific, linkage can then be found when the two treaties which are used to govern management of whales and elephants in international law are considered. Examination of the histories (at times congruent, at times divergent) of the two treaties helps to understand each. These histories go some way to elucidating the general development of states' understandings of conservation (sustainable use) and preservation.

Hopefully, treaties will in future be crafted with greater sophistication, to prevent the sort of impasse that has become a feature of the ICRW, particularly, and of CITES. However, given the nature of international law – with the sovereignty of states being relatively unimpugnable and there being no general arbiter with the authority to determine legality and the power to enforce compliance – it is likely to be some time before such a treaty can be created. In the meantime, probably the best that we can do is seek to understand as fully as we can the inherent 'biodiversity' of international agreements and the ways in which they exist in the 'ecosystem' of international law and politics – in a similar way to scientists seeking to understand, and increasingly managing to understand, the interdependent relationships of living things within ecosystems.

The world generally needs to move away from managing, and/or protecting, species in isolation and to move toward understanding, managing and protecting species within their ecosystems, in a holistic fashion. If law is to fulfil its role as a tool to achieve this, then the true nature of international law as an 'ecosystem' must be recognised. Ultimately, it is to be hoped that a framework treaty – specifically the Convention on Biological Diversity (CBD), of current treaties – might be used to broaden governance of the use of natural resources globally. The advantage of this would be that it would enable the world to move away from its present focus on single-species (and single-issue) treaties and to move toward more sophisticated governance techniques. Arguably, the world's focus on governing species singly is one of the biggest problems which we have in protecting biodiversity.

There are numerous physical environmental problems which currently bedevil efforts to conserve wildlife – such as, in no particular order, habitat destruction, the impacts of pollution and climate change, invasive alien species, the isolation of biodiversity 'islands' and the wildlife trade (both legal and illegal). There are also numerous legal problems – such as, in no particular order, the over-emphasis in international law and legal relations on the sanctity of sovereignty, the lack of effective monitoring mechanisms and the inability of those bodies which *have* been created to oversee treaty operations to enforce treaty requirements. It is to this latter list that the single-species/single-issue problem belongs, it is suggested.

As we move, in all spheres of environmental governance, toward more integrated and holistic management techniques, so international law needs to follow suit. At a time, in fact, when the world's states seem to be moving toward synergies and clusterings of international instruments, almost to a 'biodiversity of conventions', it sometimes seems that nobody wishes to touch the ICRW, and the whaling issue generally, 'with somebody else's ten-foot barge pole'. It seems, even, that there is at least a degree of careful avoidance of the whaling issue-area in other, newer, multilateral environmental agreements. One possible reason for this is that states attach so much importance to whaling that they can only get agreement elsewhere by excluding the issue-area. Another possible reason is that states are so worried by the possibility of contaminating newer treaties with the conflict that has marked the ICRW for decades that they exclude the issue-area.

CITES improves on the ICRW, but remains by its very nature (vide its approach of providing species with differing levels of protection, by listing species in Appendices) a treaty that categorises species. CITES singles out species for protection and thus may contribute to the problem that protection is not accorded to the ecosystems in which these species live or to other species which live interrelated with the protected species. One of the biggest problems which CITES faces is the difficulty of distinguishing between endangered and non-endangered species in international trade. To expect a busy, poorly trained Customs official to be able to identify a nondescript, or even dyed, bird as an endangered black-footed bog-trotter, when the person transporting the species claims that it is a non-endangered brown-footed swamp-trotter is unrealistic. Adding to this are the problems of 'look alike' species and variations within species; birds, for instance, may have female, male, juvenile, intermediate, regional and melanistic differences. Another problem lies in the trade of non-endangered species where the removal of individuals of these species might have unforeseen, perhaps even unforeseeable, impacts on endangered species which share their ecosystems – and on the general health of such ecosystems. (Remember, CITES has almost 35,000 listed species!) What is needed, therefore, is an approach which focuses on protecting habitats, ecosystems and wildlife generally, rather than the current approach which requires specialised identification.

2 Multilateral environmental treaties

2.1 Environmental treaties in the modern world

The United Nations Charter contains no direct environmental provisions. However, it is under the aegis of the United Nations that many of the world's international environmental treaties exist and derive their authority. Yet there is no body which can force a state to act in a particular way; or even to adhere to the obligations it may have agreed to in a treaty.

One of the first questions asked about international law – and a surprisingly tricky question on which to give a definitive answer – is whether international law really *is* law. Certainly it is not law in the sense in which most citizens come to understand law during their process of becoming politically aware. According to Sands, the process of negotiation of an international agreement is 'not generally understood' as it 'does not resemble law-making in national bodies' and 'the processes of making international treaties bear very little relation to what is described in textbooks' (Sands, 2006/5: 80–1). It is difficult to imagine a state in which all citizens are not merely equal to each other but to the state and its institutions also, and where there is no acknowledged legislature, judiciary or executive; yet this *is* the international legal system. It has been suggested that in any international dispute, each side 'will doubtless claim legal justification for its actions and within the international system there is no independent institution able to determine the issue and give a final decision' (Shaw, 2003: 2).

It is important to understand the process by which multilateral environmental agreements come into existence. Awareness of an environmental problem with a cross-border nature might begin at as low a level as the 'local' – within a group, village, community or municipality. Awareness might then reach the government of the country in which the process begins. Eventually, awareness might reach the international community of states, governments, itself. More cynically, the pressure from and warnings by individuals and groups might be borne out by an environmental disaster. At some point, the idea of holding an international convention might be mooted. The aspiration will be to bring together a large number of states, including relevant and powerful actor states, and to see a binding treaty result – to see the creation of binding international law, with meaningful commitments toward solving the perceived problem. Almost from the same moment, however, the 'watering down' process will begin – the hard reality of international law and politics being that states rarely commit themselves to treaties unless the commitments they make are largely ineffectual, or their bargaining strength is weak and they have acted against their own interests for the sake of gaining a different advantage. Altruism is rare in international law.

The 'watering down' aspect will almost inevitably see the result of a successful convention being not binding 'hard' law, but rather guidelines, intentions, statements of intent and aspirations – so-called 'soft' law. The best, probably, that the international environmental lawyer can hope for is that over time, as these guidelines are repeated in successive conventions and international legal instruments, they might come to harden into binding international customary law.

It may be that it will sometimes be impossible, and it might even be undesirable, to have detailed and precise rules of international environmental law – this is because 'a certain flexibility is often the price which has to be paid to secure international agreement'. The 'undesirability' aspect comes in because 'a treaty with precise rules might be hard to renegotiate' and thus be 'too inflexible to respond to changing conditions', particularly when it is environmental problems, which 'tend to require flexible solutions to allow for changing scientific evidence, new control technologies, new political priorities, and the differing circumstances of various states' that are concerned (Birnie and Boyle, 2002: 8–9).

The ICRW was agreed to in 1946, and came into force in 1949, but can arguably be said not to be 'under' the United Nations as its origins lay not in the changing world order after the Second World War, but – almost in opposition to this changing order – in the small club of nations which were actively involved in commercial whaling. Even before the War, there had been several earlier treaties on whaling which prefigured the ICRW. In some ways, the ICRW does not share the characteristics of international environmental treaties negotiated later under United Nations auspices. In fact, some analysts and states party argue that it is not an environmental treaty at all, but a trade treaty which is being used, or misused, as an environmental instrument. This argument goes too far. Although its *raison d'être* certainly lay in trade, or perhaps in control of trade, the means chosen to control or regulate trade took the form of environmental regulation.

Modern international environmental treaties are changing shape. Unlike older treaties, modern treaties more regularly include provisions for aid (financial, or in some other form) to be provided to states party – especially developing countries – when they are asked to act in ways that benefit the developed world or other states, at a cost to themselves. There is also a trend for modern treaties to include monitoring and reporting procedures, with support structures and institutional frameworks. Importantly, there is a trend toward including compliance regimes. Further, modern treaties appear to be taking more holistic approaches than in the past. The effectiveness, or lack thereof, of older treaties is worth examining in the light of changing treaty terms and practices. There are important lessons to be drawn from loopholes which become apparent in older treaties – and in the ways in which states exploit weaknesses. If the goal of international environmental law is, as it surely must be, protection of the environment and corresponding protection of human interests, then effective treaties are required – and weaknesses need to be identified and eliminated, a process that will almost inevitably require the accommodation of disparate, and often contradictory, interests.

In considering the nature of multilateral environmental agreements (MEAs), it becomes apparent that it is extraordinarily difficult to create and then maintain functional treaties able to 'stand the test of time'. The ICRW is arguably showing itself to be unable to cope with at least some of the ways in which international views on management of natural resources have changed. MEAs that are not originally designed with sufficient flexibility are ill-suited to having the views of their parties change – sometimes even to the extent that their parties hold positions the opposite of those with which they began. This would not be a problem if all parties were to change their minds in similar directions, but where some parties insist that the MEA continues to function as originally intended, and others to insist that it has changed or must change, conflict arises. In the case of the ICRW, the conflict has become so great that it is almost impossible to conceive of replacing the MEA – it would be too difficult to get agreement on a future treaty.

It does appear that Japan and Norway have attempted to use CITES in order to circumvent[1] the moratorium on whaling imposed by the International Whaling Commission (IWC) – the management body created by the ICRW. Allied to this is the potential for these, and other, states to use the international trade in ivory as a symbol and precedent for efforts to resume whaling; and for opponents of commercial whaling to use the ivory trade to support the preservationist idea in the conservation of whales. These, of course, are only two ideas. There are linkages generally between the two elderly, and yet high profile, treaties. The linkages are not only inherent, but have been made specifically by different parties themselves. The ICRW deals only with certain cetaceans, CITES deals with many species – but on occasion it has seemed, so bitter have the arguments been, as though CITES is dealing only, or mainly, with elephants (the CITES logo is even in the shape of an elephant!).

It may be contended that neither the whaling nor the ivory trading debate/issue can be fully understood in isolation from each other. Nor, ultimately, can any area of global wildlife management be fully understood without an understanding of both the whaling and ivory areas. This is because the two species are the 'global icons' of the world's battle to define the meaning of 'conservation' – and to decide on the value which will be ascribed to biological diversity ('biodiversity'), and on how this can best be protected. Neither the important political actors loosely described as adhering to a 'protectionist' philosophy, nor those of equal importance adhering to a 'sustainable use' approach, will ever be found claiming to be

1 According to Nicola Grandy, then Secretary to the International Whaling Commission (IWC), 'circumvent' is not the right word to use – suggesting that the CITES deals only with trade and cannot be used to circumvent the IWC, only to 'weaken' it (Grandy, personal communication, 2007). I have continued to use the word 'circumvent' as this would be the effect were proposals at the CITES to downlist whales to be successful.

destroying, or wishing to destroy, biodiversity. Both sides claim that biodiversity must be protected and that *their* approach is the correct one for achieving this goal.

The argument being made is that this battleground is to be found in the debates over whaling and the ivory trade. However, to my knowledge, the specific linkage of the two has not been explored. In a collection of chapters entitled *Elephants and Whales: Resources for Whom?* (Freeman and Kreuter, 1994), for example, not one of the 17 papers explores specific linkages between the two species in international law – each deals either with one species or the other. In the middle of 1997, I was in Harare and became peripherally involved with the CITES Conference of Parties. That year, the Conference was dominated by the efforts of Southern African states (ultimately partially successful) to have the African elephant downlisted from Appendix I to Appendix II, thereby allowing limited trade in ivory to resume. Two items which did not garner great media attention struck me while listening to various debates and reading news reports. The first was that the two developed countries which argued most fervently in support of the Southern African right to 'sustainable use of their sovereign natural resources' were Norway and Japan. The second was that a proposal by Norway to downlist a subpopulation of minke whales from Appendix I to Appendix II received a majority of 'yes' votes in a secret ballot – and the number of countries which voted 'yes' was greater than the entire membership of the IWC at the time. From those two seeds my Doctoral research and then this book grew.

In studying the linkages between the species in international law, it is important to understand the two treaties under which the trade in each species is regulated. To be learned from this is that international environmental treaties may have 'shelf-lives' and CITES and the ICRW might well be nearing the ends of theirs. The question is: what happens to international environmental treaties/MEAs, when they have been running for decades and the world in which their original terms were formulated has changed; the attitudes of their parties, or some of them, and of the world's states generally have changed; and they end up being used for purposes for which they were not originally intended? This is problematic for a number of reasons, including that the treaties themselves may not be suited for the purposes to which they are being applied; but parties are vulnerable to the charge that they are even acting ultra vires if they have deviated from the treaties' original purposes.

This leads necessarily to discussion of a related issue: that of differing perceptions of conservation held by different international actors. The principal actors involved in the issues hold markedly different ideas about conservation, the utilisation of natural resources and indeed the very concept of sustainable development. The problem then is that when different parties, states, hold radically different perceptions and/or understandings of key concepts, and treaties become deadlocked between these different interpretations, there seems no way forward. It is inconceivable that a deadlocked treaty can simply be replaced by a freshly negotiated treaty which will accommodate all interests – parties become so locked into their opposed positions that the negotiation of a new treaty becomes impossible. Parties therefore become locked not only into their understandings, but into the treaty itself – for myriad reasons, legal, political and economic – and become as reluctant to break away completely from the treaty as to abandon their positions within the treaty. We therefore limp on with what we have. This is a significant international law problem.

The ICRW is an MEA that often seems to stand on its own – other more recent MEAs tend not to be 'single issue' treaties. The question then arises whether the ICRW ought to be, if it can be, brought under the 'umbrella' of a wider conservation regime. If it is not, then one might ask why the ICRW is not simply scrapped or left to become dysfunctional; but it might be argued that the real importance of the ICRW lies in the reluctance of states either to leave it, or to reform it. Its importance lies in its status as a battlefield. What, then, is being fought for? It may be contended that the prize is nothing less than the determination of the philosophy of human use of natural resources which will eventually be adopted by the majority of states.

It has been argued that conservation of our terrestrial or marine wildlife resources is interpreted in different ways by different societies and cultures, with, for some, 'it meaning not to use at all (this concept being called protectionism)' and, for others, it meaning 'full-scale use' (Lapointe, 2003: 1). On the historical development of the clash between the two concepts, Thomson writes that Theodore Roosevelt conceived conservation to mean 'wise use' of 'renewable natural resources', whereupon the term preservation began to be used to mean 'protection from harm' and that, until that time, 'preservation was a rather vague term covering the gamut of wildlife management ideals'. Thomson then argues that 'each concept is distinct and true with its own function and purpose'; conservation implying that renewable natural resources can be 'used' by man, with it being implied that such use will be 'wise'; and preservation implying 'protection from harm (in the short term or the long term), which makes preservation the antithesis of conservation' (Thomson, 1986: xxv–xxvi, 25–6).

Thomson continues, suggesting that 'the two sister concepts are only readily distinct at the extremes of their respective ranges', being 'therefore part of a philosophical continuum'. He suggests, however, that it is important to draw a conceptual distinction between the two as each contains specific ideas for wildlife management, which ideas might become ineffective if blurred (ibid.: 39). According to M'Gonigle, however, it is vital that eventually the two *are* blurred, or at least brought together. He writes that 'the economic and the ecological approaches to human activity seem to present an irreconcilable conflict; we are caught between two competing paradigms', and that only if the 'nature of the conflict' is recognised and 'a resolution sought might a lasting stability for man in his environment be achieved' (M'Gonigle, 1980: 120). Miyaoka suggests that Japan accepts the 'conservation principle (sustainable wildlife utilisation) to contest the wildlife preservation principle' (Miyaoka, 2004: 118).

One of the reasons that this is such a difficult problem to resolve has to be because of the way the debate is framed: neither side's argument is in itself immoral. The labels are well chosen as it is hard to object to either – both are worthy goals. In the same sense, at a philosophical level, neither being 'pro-sustainable use' nor 'preservationist' can be impeached. The 'devil lies in the detail' and it is here that critics of either approach must base their attacks on the consequences/implications of their targets.

The word 'conservation' is, of course, a tricky one, as it is used by both 'preservationists' and 'sustainable users'. Wijnstekers, former Secretary-General of CITES, for instance, suggested that CITES is an adaptable treaty which 'has learned to balance conservation and sustainable use' (Wijnstekers, 2007). By and large, in this book conservation will be taken, however, to mean much the same as 'sustainable use' unless it is specifically or by context indicated that this is not the meaning intended (usage can be tricky, as seen in 2003 when pro-whaling states in the IWC voted against creation of a 'Conservation Committee').

2.2 Environmental treaties generally

The traditional sources of international law, according to Article 38(1) of the Statute of the International Court of Justice (ICJ), are treaties (although technically only contractual in nature), *ius cogens*, custom and the writings of jurists; the cynical commentator might suggest adding the environmental disaster (it being quite staggering sometimes how quickly states will move to create international law in the aftermath of a disaster).

All international agreements, it has been suggested, inevitably result from a compromise of a variety of national interests to achieve greater national benefit, 'as perceived by the states sacrificing these interests, accruing from international agreements restricting the interests of others as well as themselves'; and further, political compromises must be negotiated if international agreement is to be reached in whatever forum debate takes place (Birnie, 1985: 203–4). In the particular case of the IWC, however, Birnie suggests that the United Nations and its specialised agencies took no role in its foundation or subsequent development and that the history of the IWC might well have been very different had this not been the case (Birnie, 1985). This is a useful argument. The IWC is a body that has stayed out of the mainstream of MEAs, with its concentration on a single set of species, and this isolation may well have contributed to its arguably dysfunctional nature.

The counter argument to the charge that states do purely as they please is that generally states do tend to obey the commitments which they make. Lyster is optimistic, suggesting that treaties are often better enforced than is realised; partly because of improved administrative measures, but also because states, by and large, do have interests in preserving order and preventing chaos – suggesting that this is particularly true for wildlife treaties, where general system collapse would harm all participants (Lyster, 1985). It might be that this attitude on the part of states leads to them acting, on occasion, to their own detriment in favour of the common good of keeping the international legal system stable. However, it does not prevent conflict and would tend to fix in place, rather than to mend, problems – in other words, to maintain the status quo even where this is not ideal and causes hardship and dissatisfaction. It might lead also to states seeking private ways to circumvent or even deliberately undermine treaties. Further, Lyster's mid-1980s optimism might be misplaced.

It has been suggested that generally the law according to which a particular organisation operates will flow 'basically from conventional law, namely the constitution of that organization', and the practice of organisations will then be based on 'legal opinions of the legal advisers of the organizations and decisions taken by their organs will especially fill out or even expand constitutional texts', with customary international law and general principles of law being relevant when texts are required to be interpreted. The 'starting point' for interpretation of treaties, Amerasinghe argues, is the Vienna Convention on the Law of Treaties, 1969, which provides (in Article 31) that a treaty must be 'interpreted in good faith in accordance with the ordinary meaning to be given to the terms of the treaty in their context and in the light of its object and purpose'; per Article 32, supplementary means of interpretation may also be used. Amerasinghe adds that the Vienna Convention gives subsequent practice 'a substantive place in the ascertainment of the ordinary and natural meaning: subsequent practice may be taken into account in establishing such meaning' (Amerasinghe, 2005: 20–1, 41, 49). The importance of this for consideration of the ICRW, particularly, is that both those parties which argue that the treaty is not currently being interpreted as it ought to be, per its original text, and those parties which argue that it is possible for the purpose of the treaty to have changed, have cases that may not be dismissed out of hand.

2.3 The *Bering Sea Fur Seals Arbitration*

At least some of the problems faced by modern fisheries originate from 'concepts and doctrines of the law of the sea attuned to the outdated interests of earlier centuries' (Birnie and Boyle, 2002: 648–9). An account, therefore, of the seminal *Bering Sea Fur Seals Arbitration* is essential for understanding the 'open access' nature of the high seas – and therefore for understanding the whaling debate. It has been suggested that the modern rules of international environmental law can be traced back 'to the late nineteenth century, and an obscure spat between the United States and Britain' with the 'world's first reported environmental dispute concerning the little-known fur seals' (Sands, 2006/5: 71–4). At that time, international law was generally considered to allow countries territorial sovereignty over coastal water up to 3 nautical miles off their coasts – the so-called 'cannon-shot' rule of distance and control. Beyond this limit, utilisation of living resources was free to all.

In the late 1800s, fur seals were declining rapidly in the Bering Sea – being hunted by a number of countries: Britain (Canada), Japan, Russia and the US. The US argued that it had a right to protect, and of property in, the fur seals which lived parts of their life cycles on the (US) Pribilof Islands – even where these seals were temporarily outside the 3-mile limit of the US territorial sea (Birnie and Boyle, 2002). While inside the 3-mile limit, it was not disputed that the US could protect the seals. However, following its annexation of Alaska, ceded by Russia in 1867, the US claimed exclusive jurisdiction over the seals, allegedly for conservation purposes. Foreign vessels ignored the claim and continued to take seals beyond the American 3-mile limit (Birnie, 1985: 94). The US decided to arrest British ships sealing on the high seas. It seems that in 1886 the US seized three British schooners which were sealing some 60 miles from Alaska. After Britain protested, the US released the vessels; but after other such incidents, the matter was, in 1892, put for decision to a seven-member arbitral tribunal (two members were nominated by the US and two by Britain, while the President of France, the King of Italy and the King of Sweden and Norway nominated one each). The US claimed that it had, in international law, both a property right in the seals, as well as the right to protect them for the benefit of humankind generally. In support, the US invoked 'the practice of nations, the laws of natural history and the common interests of mankind' arguing that it alone had the power to preserve the seals, as a trustee 'for the benefit of mankind' (Sands, 2006/5: 71–4).

In Birnie and Boyle's words, the US regarded itself as 'the trustee of the herd for the benefit of mankind'. Britain (for Canada), however, countered that it had the right to hunt seals on the high seas; seals being 'either *res communis* or *res nullius* in status, not the exclusive property of the US'. The US countered to this that the high seas were 'free only for innocent and inoffensive use, not injurious to the just interests of any nation which borders upon it', and also that the seals had an *animus revertendi*, returning cyclically to US territory, and were thus to be equated to domestic animals which could be the subject of property rights. The arbitral tribunal, continue Birnie and Boyle, ruled (with two arbitrators dissenting) that the US had not acquired from Russia 'exclusive rights in the Bering Sea beyond areas of national jurisdiction' and so 'had no property rights in the seals and no right to protect them beyond the three-mile limit. Freedom of the high seas was held to be the prevailing doctrine' (Birnie and Boyle 2002: 649).

The arbitral tribunal's 1893 ruling is not the end of the importance of the matter. Britain and the US had agreed beforehand that if the US lost the case the arbitral tribunal should suggest new international rules to conserve the seals. The arbitral tribunal, therefore, proposed arguably the first rules of modern international conservation law, regulating when and where seals could be captured. This revealed, Sands argues, an 'American desire to put conservation above economic interests. It also reflected a willingness on the part of both countries to restrict traditional sovereign freedoms with new rules of international law' (Sands, 2006/5: 71–4). It must be, however, that Sands' contention that the Americans 'desired to put conservation above economic interests' puts American philosophy (philanthropy?) far too strongly. The US was surely protecting a resource for itself. The fur seals represented an important economic resource, and one over which the US felt it had a special proprietary interest. Support for my contention of American cynicism can perhaps be found in the fact that after losing the arbitral panel's decision, the US went on (in an arbitration with Russia in 1902) to use the same arguments as the British had put forward, after Russia had seized American vessels fishing outside of the Russian territorial sea (Joy, 2001).

The importance of this decision to the legal development of conservation of marine living resources is emphasised by Birnie and Boyle, who write that the decision 'laid the twin foundations for subsequent developments over the next century' in the sense that, first,

> it confirmed that the law was based on high seas freedom of fishing and that no distinction was to be made in this respect between fisheries and marine mammals despite the very different characteristics of the latter, which the tribunal had examined.

And second, that the decision 'recognized the need for conservation to prevent over-exploitation and decline of a hunted species, but because of the former finding, it made this dependent on the express acceptance of regulation by participants in the fishery' (Birnie and Boyle, 2002: 649–50). The regulations suggested by the arbitral panel, to be binding on the US and Britain, included that there should be no capture or pursuit of fur seals within a 60-mile zone around the Pribilof Islands, and that there be a closed season between 1 May and 31 July.

It is interesting to see this arbitration as having been an early model for the ICRW – or at least as an important stepping stone toward that agreement. The eventual 'moratorium' on commercial whaling agreed to by the IWC in 1982 being, perhaps, the most important eventual consequence. Generally, the history of multilateral conservation agreements has, though, seen the international community moving slowly and painfully toward better understanding.

The importance of the award lay in its international stature. There had been earlier moves toward some of the conservation provisions suggested. The close season, for instance, had been suggested in the early 1870s in regard to seals in Greenland. At the time, approximately 80,000 saddle back and hooded seals were being slaughtered annually. The death toll was much higher, as many pups were left orphaned. Apparently as early as 1876 an international closed period had been established prohibiting, from 1877, the killing of seals before 3 April in a year (Sutherland, 1993). These moves did not receive the attention, nor provide the precedent value, that the *Arbitration* did.

In the result, however, the legacy of the *Arbitration* must be seen as having been mixed. It entrenched freedom of the high seas and thereby made it that much more difficult to bring an end to the rampant destruction of whale populations, and left us to this day labouring under the weight of the impression many important actors have: that they can do as they please with natural resources on the high seas. At the same time, many of the principal 'tools' we use even today to give content to management systems protective of wildlife have their origins in the arbitral award. These 'tools' particularly include closed seasons and sanctuaries.

In the short term, the *Bering Sea Fur Seals Arbitration* did not meaningfully assist the fur seals themselves. This could hardly be otherwise, as it was essentially nothing more than a contract between two countries and these were only two of the four states engaged in hunting the Bering Sea fur seals; Russia and Japan being the others. In fact, some US and Canadian vessel owners apparently re-registered their vessels under Japanese and other flags to evade the US and Canadian regulations. The dramatic decline in seal stocks in the area continued until it was eventually realised by all the participants that only concluding an international regulatory treaty among all states involved in the sealing could save them. This cycle of events, comment Birnie and Boyle, 'has been repeated in almost all exploited fisheries' (2002: 650). Unfortunately, this cycle of overexploitation has yet to be recognised on the global scale, and in this sense we have not learned all that we could from the arbitral award. Dahmani comments that the freedom of the high seas was established as a justification for 'any use of the seas' and that where fisheries were concerned, 'as long as those resources were believed to be inexhaustible, no form of regulation or restriction upon the freedom of the seas could be justified' (Dahmani, 1987: 1–2).

Ultimately, in the absence of a substantial body of international case law, commentators must make much of the few cases – or arbitrations – that do exist. The precedent value of the *Bering Sea Fur Seals Arbitration* is immense, as it set the scene for the accepted 'open access' status of the high seas. The scene was therefore set for lack of control over fisheries generally – which lack of control persists – and over whaling – which resulted in incredible damage being done to most of the great whales, the creation of the ICRW, the eventual 'moratorium' on commercial whaling and the current controversy. The arbitral ruling therefore provides a useful starting point for understanding the histories of whaling and of fishing.

2.4 Early conventions and categorisation of species

Conservation measures in British and German territories in the 1890s, and cooperation between the British and the Germans, culminated in the Convention for the Preservation of Wild Animals, Birds and Fish in Africa, signed in London in 1900.[2] Most parties never ratified the Convention, but the British and Germans did so 'enthusiastically' (Mackenzie, 1988: 208). It was important for certain African countries, British and German colonies in particular, because Britain and Germany attempted to give it expression in their colonies. The agreement aimed at preventing uncontrolled massacres of wild animals and ensuring the conservation of diverse wild animal species (IUCN, 2006). It set up a selected mechanism for the protection of 'useful' or rare and endangered wild animal species and the sufficient reduction of 'pest' species (Articles II(1), II(13) and II(15)). It also encouraged signatories to engage in the creation of 'reserves' (Article II(5)). The Convention never entered into force. It is unknown whether this was because its provisions were too strict for its parties or because of undisclosed political reasons. However, this first initiative was not worthless. It helped some signatories to enact legislation related to the protection of wild fauna in their respective colonial territories (IUCN, 2006).

There were aspects of the Convention which were extremely protective, such as that it listed animals which could not be killed in any circumstances, ordained protection of female and immature animals, ordered creation of reserves and called for strict regulation of trade, amongst others. However, the Convention contained

2 The Convention was signed by the Congo, France, Germany, Great Britain, Italy, Portugal and Spain.

a clause which encouraged 'measures for effecting the sufficient reduction of the numbers of the animals of the species mentioned in Schedule V' (Art. II.13) – these being 'lion, leopard, hyena, hunting dog, otter, baboons, large birds of prey, crocodiles, poisonous snakes, pythons' (Sch. V), many of which are today seen as requiring stringent protection. Conservationists worldwide, certainly as reflected in international conventions currently in force, still categorise species.

The Convention for the Protection of Birds Useful to Agriculture, signed in Paris in 1902 and which came into force in 1905, categorised certain birds as being useful to man for agricultural purposes, and categorised certain other bird species as 'noxious'. This was done largely on the basis of which species were seen as being useful because they were insectivorous and therefore able to assist man in the protection of crops, and those which were seen as competing with man's interests or competing with bird species useful to man. Ironically, many of these 'noxious' birds are those to which we today give greatest protection.

Article I of the 1902 Convention provided that birds useful to agriculture,

> particularly the insect-eaters and namely those birds enumerated in the first Schedule attached, shall be unconditionally protected by a prohibition forbidding them to be killed in any way whatsoever, as well as the destruction of their nests, eggs and broods.
>
> (Article I)

Schedule I is entitled 'Useful birds'. It is Schedule II, entitled 'Noxious birds', which makes for staggering reading for the modern environmentalist. Consider the following list of bird species which parties were encouraged to exterminate:

> Birds of prey: bearded vulture; eagles, all kinds of; sea eagles, all kinds of; osprey; kites, black-shouldered kites, swallowtailed kites; falcons, gyr-falcons, peregrine falcons, hobby, merlin-stone falcon, all sorts of; common goose-hawk; harriers. Owls: eagle owl. Perching-birds: raven; magpie; common jay. Herons: grey and purple herons; bittern; night heron. Swimming birds: pelican; cormorant; smews; divers.

Ultimately, the 1902 Convention was superseded in 1950 by the International Convention for the Protection of Birds. This latter treaty moved beyond regulating the killing, capturing and trading of birds and encouraged its parties to promote conservation education and establish protected areas (Glavovic, 1995). This second Paris Convention, which entered into force in 1963, suggested in its preamble that its signatories considered that 'in the interests of science, the protection of nature and the economy of each nation, all birds should as a matter of principle be protected'.[3] Two important shifts were represented by this: these being in

3 The Convention was signed by Austria, Belgium, Bulgaria, France, Greece, Monaco, the Netherlands, Portugal, Spain, Sweden, Switzerland and Turkey.

perception of the value of birds, and in focus from species protection to habitat protection. It has taken humankind a very long time to reach our current understanding of the complexities of species interrelationships and it would be dangerously arrogant to make the assumption that our current understanding is correct. Probably at the time of the drafting of the first Paris Convention, scientists had similar confidence in their understanding to that which scientists have today. What understanding we will eventually come to have is impossible to know, and caution is essential.

Following the failure of the London Convention of 1900, in 1933 the Convention Relative to the Preservation of Fauna and Flora in their Natural State was adopted. This new London Convention came into force in 1936. Compared to its predecessor, the scope of the Convention was extended markedly – even to include plants. Its objective remained utilitarian – to preserve supplies of species which were economically valuable and popular with trophy hunters. In this regard, it provides a list of plant and animal species that were selected either to receive absolute protection (class A) or a lower level of protection (class B). The Convention, however, took a bold step into the future by rejecting the concept of nuisance species. It was also the first binding instrument, binding on the states which adopted it at least, to provide for the creation of protected areas in Africa.[4]

In 1946, 15 whaling states created the ICRW with its managing body, the IWC, which would meet annually to determine quotas for the parties – which quotas the parties would then, for decades, largely ignore. As well as setting quotas within cetacean species, the parties, through the IWC, listed the species to which the Convention would apply.[5] The Convention does not apply to so-called 'small cetaceans' (such as dolphins, pilot whales, porpoises, etc.). 'Small cetacean' is not a strictly biological term, and does not appear in the Convention or its Schedule, but is rather a later informal coinage – where the legal basis for excluding species is merely that they were not initially listed, this must be considered as, at best, spurious (Mulvaney and McKay, 2003). It may even be that the Baird's beaked whale, for instance, was omitted from the list because it was simply not known to Western nations – being hunted in Japan only (Komatsu and Misaki, 2003). What this categorisation of species has led to is a situation where certain species are

4 The Convention was signed by Belgium, Egypt, France, Italy, Portugal, South Africa, Spain, Sudan and the UK; India and Tanzania later acceded.

5 In the 'Annex of Nomenclature of Whales', which the contracting governments agreed to in 1946, approximately 'a dozen' (IWC, 2013) species of whale are listed. The Annex is difficult to read. Species are not numbered, the order is haphazard and many of the scientific names have changed. Some of the species have more than one name – the gray whale, for instance, has eight (gray whale, gray back, California gray whale, Pacific gray whale, Devil fish, hard head, mussel digger, rip sack) – and they are listed here according to the name I have taken to be presently the most commonly used, in the order they appear on the Annex: bowhead, right whale, North Atlantic right whale, Southern right whale, pygmy right whale, humpback whale, blue whale, fin whale, sei whale, Bryde's whale, minke whale, gray whale, sperm whale, Arctic bottlenose whale and Antarctic bottlenose whale (IWC 'Annex: Nomenclature of Whales', 1946).

hunted, and others not. As an example of the arbitrary nature of such selection, the northern bottlenose whale is not considered a 'small cetacean'; while the larger Baird's beaked whale *is* so considered (Mulvaney and McKay, 2003).

The hunting of small cetaceans is one of the more controversial aspects of the regulation of whaling by the IWC. While the matter is not officially under the auspices of the IWC, according to the terms of the ICRW, member states opposed to the hunting of small cetaceans have often succeeded in having the matter put on the agenda for discussion at annual meetings. While there is no way, in international law, for those opposed to the practice to stop it, they do put public pressure on those countries – such as Japan and the Faroes – which do practise such hunting.

As will become apparent, CITES is ostensibly a protectionist treaty – and quite firmly rooted in the emerging environmental discourse of the early 1970s. However, the implicit recognition that trade is a reality (and is 'here to stay') leads one to the inevitable conclusion that the treaty is also a trade treaty – a 'regulation of trade' treaty. This is a problematic approach for a treaty that is intended to be protective of species and, ultimately, of biological diversity. The treaty takes a fairly rigid approach toward categorising species, and yet, having only extremely limited internal jurisdiction, does not acknowledge that species might depend upon other species, in properly functioning ecosystems, for survival. There are numerous related problems, such as that of so-called 'lookalike' species – where species in desperate need of protection might not receive it due to being confused with species not in such need.

Article II of CITES is headed 'Fundamental Principles' and it creates the three Appendices into which listed species are placed and in terms of which they are thereby afforded different degrees of protection against overexploitation in trade. Article II.1 states that Appendix I species include all species threatened with extinction which are or may be affected by trade. Trade in these specimens must be subject to particularly strict regulation in order not to endanger their survival further and trade must only be authorised in exceptional circumstances. Article II.2 states that Appendix II includes all species which might become threatened with extinction unless trade in specimens of such species is subject to strict regulation 'in order to avoid utilisation incompatible with their survival'; and also other species which 'must be subject to regulation in order that trade in specimens of certain species may be brought under effective control'. Article II.3 states that Appendix III includes 'all species which any Party identifies as being subject to regulation within its jurisdiction for the purpose of preventing of restricting exploitation, and as needing the cooperation of other Parties in the control of trade'.

Arising from historical consideration and a necessarily brief look at the ways in which the ICRW and CITES operate, it is apparent that the idea of categorisation of species is dominant. The protection of wildlife species has historically operated on a strict system of classification according to perceived levels of danger to populations of particular species. The London Convention of 1900 and the Paris Convention of 1902 both make rigid distinctions between species – setting the scene for the approach adopted by both the ICRW and CITES. This approach arguably makes it extremely difficult to use elderly MEAs to protect species and biological diversity generally – in the face of our current, much improved, understanding of intra- and inter-species relationships, ecosystems and biodiversity.

3 The first old watchdog

The ICRW

3.1 The early history

As an interesting aside with which to begin discussing the ICRW and CITES, both can be referred to as 'the Washington Convention' as this is the city where each was originally adopted – in the years 1946 and 1973 respectively.

No matter how briefly described, the sheer effort inherent in whaling cannot fail to shock. Apparently, by the middle of the nineteenth century, nearly 700 American whaleships, carrying 200,000 crew, sailed annually (Creighton, 1995), and during just the first three-quarters of the nineteenth century, harvested almost 200,000 right whales and in total, about 300,000 whales of different species (bowhead, gray, right and sperm) (Nikonorov, 1996). That, of course, describes only the US! Gradually, whale populations were 'mined out' of the world's seas, much as was already happening to mammals in Africa and in the US.

There were some early efforts at regulation, in countries like Australia, Britain and Norway, which were of no effect in international waters. It seems that in 1911, the suggestion of creating an international agreement to govern whaling was made by staff of the Natural History Museum, London, with further suggestions being made in 1913 through a French proposal arising from concern over whaling off Congo (then a French territory) (Martin, 2001). Then, in 1924, the Assembly of the League of Nations initiated an inquiry into international law generally, and this led to a discussion draft on whaling based on regulations that the Norwegian government had established for Norwegian whalers – the effort being described as 'another fruitless attempt, almost impossible to police', but arguably marking another point in the 'long struggle to exert some form of management over whale stocks' (Martin, 2001: 80–3). In 1931, the Convention for the Regulation of Whaling (CRW), an agreement supported by the Assembly of the League of Nations, was opened for signature. The next important step would come in 1937 when a number of whaling nations signed the International Agreement for the Regulation of Whaling (ibid.). The real importance of these early efforts was that, as with so much international environmental law, they provided basic starting points, building blocks, rather than that they were themselves effective law.

The 1931 Convention did come into effect in 1935.[1] Although the Convention did, for the first time, attempt to extend its scope to 'all the waters of the world', it probably 'did little more than rally the support of nations in the cause of conservation' (Birnie, 1985: 117–8) and was 'at best, a very tentative step towards conservation . . . At worst, it was a complete whitewash of the problem' (Scarff, 1977: 349–50). It did, however, in its failure lead to a different form of international 'agreement' – independent regulatory efforts by the industry itself.[2]

Birnie tells us that in 1936 negotiations began toward a more effective quota agreement with governments joining the companies, and that in 1937 an international convention was negotiated, valid for one season, with a subsequent follow-up conference then convened in 1938 (Birnie, 1985: 126–7). In 1944, the conference reconvened.[3] After the Second World War, in 1945, the UK convened a conference in London with such issues as the creation of a Standing Commission, confiscation of oil taken illegally by non-party states and the forbidding of imports and sales of whale products by non-party states. These ideas presaged the regime that was eventually to be adopted under CITES (ibid.: 132–3). Already, ideas of conservation of stocks relied upon trade controls and restrictions put in place in early whaling-related international thinking – foreshadowing CITES. In this sense, then, the ICRW and CITES can be seen as having been inherently linked from their very beginnings.

The ICRW, 1946, came into force in 1948 and established the International Whaling Commission (IWC). The IWC consists of one member appointed by each ratified party to the ICRW; that member might be accompanied by alternates and other delegates. The ICRW originally came into existence to provide for 'the proper conservation of whale stocks' and thus to 'make possible the orderly development of the whaling industry' (ICRW, Preamble). The stated goal of the IWC is thus to protect all species (as listed on the 'Annex of Nomenclature of Whales', that is) of whales from overexploitation. This was not, however, a protectionist philosophy. The goal was the utilitarian one of gaining maximum benefit from the sustainable

1 The Convention was ratified by Canada, Denmark, Finland, France, Italy, Mexico, the Netherlands, New Zealand, Norway, Poland, Spain, Switzerland, South Africa, Turkey, the UK, the US and Yugoslavia, and subsequently adhered to by Brazil, Ecuador, Egypt, Eire, Great Britain (for its colonies), Latvia, Monaco, Nicaragua, Newfoundland and Sudan. Albania, Australia, Belgium, Columbia, Germany, Greece, India and Romania signed but did not ratify (Birnie, 1985).

2 Scarff explains that the 'failure of governments to achieve an international agreement led to independent efforts by the whaling companies to regulate the industry themselves' (Scarff, 1977: 350). In 1932, all major whaling companies had joined in an agreement to limit the production of whale oil. Individual companies were assigned harvest quotas based on a 'blue whale unit' (BWU) (Scarff, 1977). Initially, one blue whale was considered the equivalent of three humpback or five sei whales. Although the cartel broke down by 1934/5, the BWU was re-introduced in 1939 by the Bureau of International Whaling Statistics as a way to gauge the catches of various species (Frost, 1978: 28).

3 At this reconvening, the BWU was revised so that one blue whale came to equal two fin whales, two and a half humpbacks or six sei whales (Frost, 1978).

usage of a natural resource. The countries opposed to whaling were eventually able to place a moratorium (technically, a 'zero quota') on whaling only because sufficient countries opposed to whaling joined and the countries in favour of whaling were outnumbered; not because this was originally envisaged as an option. In fact, the possibility of such a moratorium was perhaps not even thought of by the original signatories. Historical irony is a significant feature of the IWC today: often states can be seen to have moved to positions diametrically opposed to those they held earlier. Perhaps the 'missionary zeal' of the convert is partly to blame; perhaps, more cynically, the explanation is that states truly do act, at least largely, out of self-interest. This point is not as trite as it might at first seem. In the morass of explanations which states give for their actions, it might be difficult sometimes to remember the extent to which self-interest underlies all.

Scarff suggests that the Convention has always had a serious flaw in that 'it strives to achieve several goals which have proven incompatible' (1977: 353–4). Birnie, however, argues that the IWC's problems are similar to those which confront all international bodies managing open access migratory resources. She describes the IWC as being the first commission to be established on a global scale and asserts that some of its difficulties related to this 'pioneer' role, especially given the 'constraints imposed by legal principles developed in a different age, before the development of twentieth century technology and advances in scientific knowledge' (Birnie, 1985: 143). Jacobson contends that the 'main, crystal-clear objective of the treaty was, and still is, to provide for the conservation of whales for the benefit of the whaling industry and its customers'. He contends further that the ICRW is 'one of the primary early treaty commitments to the ideal expressed today as sustainable development of the earth's natural resources' – an ideal which he says is 'heavily promoted by environmentalists in recent decades' and 'has probably emerged by now as a principle of customary international environmental law' (Jacobson, 2001: 81–3).

An interesting argument is that the ICRW in fact has turned 180 degrees and today reflects the views of environmentalists. To describe the 'ideal' of sustainable development as 'a principle of customary international environmental law' probably goes too far, however, in that sustainable development ought, in my view, to be seen as an 'overarching goal' rather than a principle. Principles (such as the precautionary principle and polluter pays principle, insofar as these have the status of principles in international law) are more specific and help to attain goals. Sustainable development would not, in any case, be a very practical principle. Jacobson is, therefore, arguably being disingenuous and conflating sustainable development with conservation (in its sense of 'wise use').

Fifteen states attended the 1946 Washington Conference – Argentina, Australia, Brazil, Canada, Chile, Denmark, France, the Netherlands, New Zealand, Norway, Peru, Union of South Africa, the UK, the US and the USSR. Iceland, Ireland, Portugal and Sweden were observers.

Birnie argues that it may actually be advantageous for the preservation of whales that the ICRW was not more specific in its original definitions, suggesting that the lack of a definition in the ICRW for the phrase 'optimum level' carries the inherent advantage that the terms remain open to a more flexible interpretation than would have been the case if the term had been defined (1985: 169–70). Her suggestion was made, of course, in the mid-1980s at a time when the validity of the preservation of whales was a relatively new idea that was still being fought for.

The original parties showed great reluctance to give up any more of their freedoms than absolutely necessary; something easily understandable in the immediate aftermath of the Second World War. Contracting governments could avoid being bound by Schedule amendments by lodging objections within 90 days thereof; according to M'Gonigle, the US had 'insisted on this provision in 1946, ever concerned not to sacrifice any national sovereignty' (1980: 185). Per Birnie, however, many strengths of the ICRW were already apparent. The reasons she gives are that the ICRW provided for a continuing supervisory body; that regulations could be incorporated not in the Convention as such, but in a flexible amendable Schedule that continued to apply, whether amended or not; and that the IWC was able to make use of its Rules of Procedure 'to broaden and strengthen without the necessity of convening a conference of plenipotentiaries' (Birnie, 1985: 201–2).

By the second meeting (IWC 2) in 1950, and IWC 3 (1951), there were 21 states present as either parties or observers, and by then there were 16 ratifications. New states to have ratified included Brazil, Denmark, Japan, Mexico and New Zealand. At this meeting it was decided that the IWC should remain outside of the ambit of any other international organisation (Birnie, 1985: 212–3). As this included the United Nations Food and Agriculture Organisation (FAO), it occurs to me that the decision was to have important consequences – a closer relationship with the FAO, which would have emphasised the food aspect of whaling, might well have made it more difficult eventually for preservation-minded parties to argue that one of the ICRW's tasks is to consider whale preservation.

In 1952, three developing countries – Chile, Ecuador and Peru – signed the Declaration of Santiago, in terms of which they claimed exclusive rights for themselves over a maritime zone 200 miles off their respective coats. At the same time, the three adopted a Joint Declaration on Fishery Problems in the South Pacific and agreed to establish the Permanent Commission of the Conference on the Use and Conservation of the Maritime Resources of the South Pacific (PCSP). 'However understandable', comments Birnie,

> because of the depredations by foreign fleets of the fishing grounds of these countries and the [1945] Truman Proclamation of [the US] claiming exclusive rights to the seabed resources of the continental shelf, the actions of these developing states threatened the IWC's effectiveness and greatly undermined the implementation of its regulations until 1979, when Chile and Peru finally joined the IWC. The PCSP argued that the IWC regime favored wealthy countries and damaged poor countries.
>
> (1985b: 941–2)

It is ironic that today the same argument is made, but the former PCSP members are now on the side of their erstwhile 'rich' opponents – and opposed to 'poor countries' which base their pro-whaling arguments on 'food security' and 'cultural imperialism' issues.

As the early annual meetings got under way, new states gradually began to attend – usually initially as observers (Italy, Panama and Portugal, for example). It also became apparent that whale stocks were continuing to decline, and much dispute centred on the issue of Antarctic whaling. IWC 10, 1958, saw a proposal introduced that those parties involved in Antarctic whaling should set their own quotas outside the auspices of the IWC – since the IWC did not provide for setting national quotas (Birnie, 1985: 216, 242).

In 1959, three parties – Japan, the Netherlands and Norway – threatened to withdraw. The issue was the failure of Antarctic whaling states to agree on catch allocations. The Netherlands and Norway then allowed their withdrawals to take effect. Japan, however, chose to withdraw its notice, stating that it 'fully recognised the exemplary and rational nature of the present Convention as a means of preserving collective operations' (Birnie, 1985: 250). This is interesting. Japan has on several occasions over the years threatened to withdraw – including recently at IWC 59 in 2007. However, it never has done so – despite its supposed isolationist stance, and despite being treated on occasion as something of a global pariah because of the whaling issue. It is a positive sign that Japan has chosen to remain within the international community, as much as possible; on the other hand, Japan has – as will become apparent – been at times both subversive and hypocritical. It is strange to speculate that if Japan had been less hardline in the late 1950s, there might perhaps never have been a moratorium, and today we might be living with, and even be fairly accustomed to, whaling. After all, it was the excesses of whaling – not whaling itself – that led to a moratorium that was not easily won.

The next meeting, IWC 12, 1960, has been described as 'difficult' – given the absence of Norway and the Netherlands (Birnie, 1985: 254). Clearly, without two of its most important members, the IWC could not function properly, showing how fragile an international convention can be, particularly in its early years. This fragility is something that should not be forgotten even today, however, as the ICRW might be an example of a convention that has never truly 'worked'.

An interesting perspective on the original application and effectiveness of the treaty can be found in the writing of Robertson in a book published in 1954. *Of Whales and Men* is an account of eight months the writer spent aboard a British whaling ship in the South Atlantic. Robertson suggests that inspectors employed by the IWC faced problems of double loyalty: to whichever government employed them, and to the Convention itself. He describes such inspectors as having 'a hundred or more regulations and restrictions to enforce' and having to be 'seamen, zoologists, mathematicians, men of absolute integrity who can refuse bribes presented daily in sundry attractive forms . . . detectives with the observation and acumen of Holmes . . . in order to impose the international law', while at the same time remaining 'on terms of good will with avaricious whale owners and with hundreds of whalemen, every one of whom is trying to hoodwink them' (Robertson, 1954: 195–6). The gentle irony here captures well the attitudes of the time toward restrictions on whaling.

Vogler argues that there was never any sense that whales should be recognised as a highly mobile 'common heritage' resource in which everyone had a legitimate interest, and that the Convention was an 'arrangement' between commercial exploiters (Vogler, 2000). Lyster agrees only partially, describing the Convention as 'by no means a protectionist treaty', and explaining that the phrase 'and thus make possible the orderly development of the whaling industry' makes it clear that the main object of the Convention is conservation of whale stocks for the secondary object of enabling the whaling industry to 'develop in an orderly fashion' – with conservation being the 'top priority' and development of the industry coming next (Lyster, 1985: 20). Whatever the true original emphasis of the treaty may have been, what is not in doubt is that in its very early years its parties acted as though Vogler's interpretation were the correct one. This does not mean, though, that this makes it the true interpretation, although it is of course a good indication; it is possible that parties acted in their own interest in ways that they knew were contrary to the import of the treaty, relying on the ambiguity that necessarily surrounded the early interpretation of so contentious a treaty.

Day suggests that the IWC was a 'whalers' club', stating that under the direction of the IWC a greater number of whales was killed than had ever before been the case. He ascribes this to there being no proper monitoring system to detect infractions, untrustworthy observers, a lack of punishment for transgressors and the fact that IWC members often violated their own set quotas, if these did not suit them, by simply filing an 'objection' and continuing the kill (Day, 1992: 41–3). M'Gonigle agrees with this, arguing that the establishment of seasonal time limits 'encouraged each nation to develop the most efficient vessels and killing techniques possible to maximize its share of the Antarctic quota' (1980: 137).

IWC 13, 1961, saw the Netherlands still missing; while Norway had rejoined in 1960, but gave notice of its intention to withdraw again at the end of 1961. By IWC 14, 1962, there were 18 members and the Netherlands had rejoined. Nothing really had changed, however. Whale populations, all available evidence showed, were continuing to decline, the Scientific Committee continued to advise reductions in quotas and members continued to ignore its recommendations (Birnie, 1985: 303).

Given what we know of IWC member positions in the twenty-first century, it is sometimes quite bizarre to read accounts of the early years of the Convention. Scarff writes, for instance, that between 1952 and 1962 the 'major opponent' to conservation measures was the Netherlands, despite its not being a whaling country, with its delegate to the IWC during this ten-year period relying 'adamantly' upon 'an unjustifiable biological analysis that resulted in higher population estimates than those made by the other members of the Scientific Committee' (Scarff, 1977: 361). Today, the Netherlands is one of the most protectionist parties.

Birnie describes IWC 15, 1963, as having 'presented the last chance' to halt the slide of whale populations. The opportunity, almost needless to say, was not taken. Japan's delegates, supported by those of the USSR, took the line that it was not for non-pelagic whaling countries, for whose economies the whaling industry was not important, to dictate to Japan – already, Japan was beginning to argue that the problem could not be discussed 'only in terms of conservation of whale stocks' (Birnie, 1985: 329). At the time, the word 'conservation' was clearly understood to mean 'preservation' – and not to be a goal of the whaling members. As will be seen, the debate has since then become more complicated. M'Gonigle suggests that the face of pelagic whaling was beginning to change, however, recording that in the years 1963 and 1964, the UK and the Netherlands sold their fleets and quotas to Japan and themselves withdrew from the industry – Japan being left with 52 per cent and the Soviet Union with 20 per cent of the IWC quota (1980: 139).

As with the earlier creation of the IWC, it has been argued that later policies of the IWC, such as the setting of quota units, were put in place because the main function of the IWC remained the control of whale oil prices (using the Blue Whale Unit – BWU) and was not species conservation (Komatsu and Misaki, 2003). When protective measures were taken, these were always too weak and taken too late – reminiscent of protective national wildlife-related legislation worldwide.[4] Subsequent meetings, in the late 1960s, saw familiar trends continue. Scientific evidence continued to show that whale populations were declining; agitation increased for stricter quotas, whilst existing quotas tended not to be filled; whaling members continued to argue for increased quotas; and hunting efforts moved toward less-exploited species. The BWU remained in place. New Zealand, increasingly supporting a preservationist approach, in 1968 announced its withdrawal again; in 1969 so did the Netherlands.

This devastation had, of course, been long predicted. In *Moby Dick*, Herman Melville wrote of 'almost omniscient look-outs at the mast-heads' penetrating 'the remotest secret drawers and lockers of the world' with a 'thousand harpoons and lances darted along all continental coasts', and asked whether 'Leviathan' could for much longer:

> endure so wide a chase, and so remorseless a havoc; whether he must not at last be exterminated from the waters, and the last whale, like the last man, smoke his last pipe, and then himself evaporate in the final puff.
>
> (Melville, 1851/1994: 436)

3.2 A sea change in attitude

At the meetings in 1970 and 1971, whaling members sought relaxations of existing restrictions and the BWU remained entrenched as the primary management tool – although Birnie does argue that the 1971 meeting was significant because of attention beginning to turn toward the stocks of smaller cetaceans (minke whales, for example), an example of the tendency to seek smaller targets as the larger species

4 For example, measures taken in the Cape Colony of what was to become South Africa to protect the quagga. By the time the quagga was given legislative protection in 1886, the last known specimen had died (in an Amsterdam zoo in 1883).

became reduced in number. It was significant also as some states began to argue that *all* cetaceans should be brought under the regulation of the IWC (Birnie, 1985) – a call that is still being made today, in the twenty-first century.

Shortly before IWC 24, 1972, the United Nations Conference on the Human Environment (UNCHE) met in Stockholm – at that time the largest ever gathering of Heads of State and the first ever truly global conference on environmental issues. The Conference produced the Stockholm Declaration, an important set of guiding principles, and led directly to the formation of the United Nations Environment Programme (UNEP). It also produced a resolution on whaling (recommending a ten-year moratorium on commercial whaling) which was forwarded to the IWC. Apparently it had been in response to a US proposal at the UNCHE that states unanimously adopted Resolution 33 which read:

> it is recommended that Governments agree to strengthen the [IWC], to increase international research efforts, and as a matter of urgency to call for an international agreement under the auspices of the [IWC] and involving all governments concerned, for a ten-year moratorium on commercial whaling.
>
> (M'Gonigle, 1980: 141)

At IWC 24, the BWU was abandoned – at least in respect of Antarctic catches – but the IWC's Scientific Committee considered, per Birnie, that a complete moratorium was not scientifically justified, since this would have the same foundation as the BWU quota, being an attempt 'to regulate several stocks as one group whereas prudent management requires regulation of the stocks individually' (Birnie, 1985: 422).

For the first time also, the US found itself caught in a contradictory position in respect of aboriginal subsistence whaling. The country had been one of the main drivers behind the UNCHE, and had begun to provide whales with significant protection under its domestic legislation; but at IWC 24 found itself facing calls for more stringent accounting for whales taken by its aboriginal populations (Birnie, 1985: 425–6).

Birnie considers there to have been two significant events in the early 1970s which had an impact on whaling then. One was the UNCHE. The other was the change in domestic policy in the US. In 1969, the US amended its Endangered Species Preservation Act (ESPA) of 1966, which originally enabled it to create a list of national wildlife and fish threatened with extinction. The 1969 amendments allowed for increased listing of species, to include those threatened with worldwide extinction – and including even species located within foreign jurisdictions or on the high seas – and largely prohibited importation of such species into the US. In 1972, the US adopted a more comprehensive Marine Mammal Protection Act (MMPA), which introduced a moratorium on the taking in the US of all marine mammals and products, and prohibited their importation (Birnie, 1985: 947–8).

Further, the US strengthened the ESPA with the Endangered Species Act (ESA) of 1973 by recognising the 'aesthetic, ecological, educational, historical, recreational and scientific value' of endangered species of wildlife; by providing for the conservation of these; and by 'urging all US federal agencies to help achieve these objectives by taking all measures necessary to remove species from the endangered category' (ss 2, 4). Effective sanctions were provided for, in respect of the international aspects of the above legislation, by Congressional amendment of two major national Acts dealing with fisheries. In 1978, Congress adopted the Pelly Amendment to the Fishermen's Protective Act, 1967, and in 1979 passed the Packwood-Magnuson Amendment to the Fishery Conservation and Management Act, 1976. Significantly for events to come within the decade, the amendments encouraged non-member states to join the IWC (Birnie, 1985: 948). The strategic introduction as ICRW members of states not actively engaged in whaling was to become an important feature.

The US began to introduce to the IWC, stimulated perhaps by its national Acts, this new perspective on international management of endangered species and to urge the IWC to switch to 'conservation rather than exploitation' as its main objective (Birnie, 1985: 949). The US, per Birnie, 'also proposed ecological principles of management and preservation of species habitats' (ibid.). This has to be recognised as ironic in the light of present policies. The irony deepens as Birnie continues, explaining that the US 'urged the Commission to adopt a perspective on environmental factors affecting population estimates much broader than the IWC or its Scientific Committee ever before considered'. Over the next decade, she suggests, several other states, following the lead of the US, began to raise at the IWC the issue of 'interrelationship and interdependence of whales with other species and their food chains, and to take into account when estimating stock yields the impact of pollution and the effects of current and temperature changes' ((ibid.: 949–50)). There are a number of ironies here, as will become apparent later. Today, in the twenty-first century, it is the pro-whaling countries which are proposing ecological principles and asking the IWC to consider the interrelationship and interdependence of whales with other species and their food chains – in fact, this has become one of the major arguments of those countries wishing to resume commercial whaling.

Globally, mores were beginning to change, however, and the early 1970s reflected the new environmental understanding and concern which had begun in Western countries in the 1960s. Countries began to recognise the whale as a symbol of environmental concern. However, the call from the UNCHE did not immediately appear to have made much impression on the IWC. Day explains that the IWC's response was to 'listen politely to the plea of the conservationist spokesmen', but then for only four of the 14 nations (Argentina, Mexico, the UK and the US) to support the UN moratorium in a vote. Six nations voted against the moratorium: Iceland, Japan, Norway, Panama, Russia and South Africa. Australia, Canada, Denmark and France abstained (1992: 43–9). Scarff comments that the IWC 'had been directly rebuked by the world community' (1977: 367). Of course, one might turn this around and suggest that it was the Commission that had 'rebuked' the world community!

IWC 25, 1973, saw the Scientific Committee continue to argue that there was no 'biological requirement for a blanket moratorium' on all commercial whaling, as this would be contrary to the objective of the ICRW, which implied that the resources 'should be so managed as to keep them in a condition which will enable them to provide the optimum yield on a continuing basis'. Something of interest for future developments which arose from the same meeting is that the parties (state members and non-governmental) who argued in favour of a moratorium did so largely on the basis of the 'need for an ecosystem approach' (Birnie, 1985: 434, 422–3).

At IWC 25, a proposal was considered for a worldwide moratorium on commercial whaling. The proposal was considered by the Scientific Committee and the Technical Committee, the former considering that there was no biological requirement or justification for the imposition of a blanket moratorium on all commercial whaling (which is not dissimilar to the Scientific Committee's arguments today, in the twenty-first century – in fact, the Scientific Committee has remained relatively stable in its views through the decades with it being the reactions of the parties to its advice that have varied); but in the latter a proposal by the US that:

> the Commission decide that commercial whaling for all species of cetaceans should cease for a period of ten years beginning not later than three years from the date of the adoption of the resolution, except aboriginal catches where they do not endanger the species

was considered. The proposed resolution was approved by the Committee by majority vote. In the plenary session, however, the Soviet commissioner repeated that the USSR was opposed to the resolution 'on the grounds that it was contrary to the findings of the Scientific Committee, that it was incompatible with the Convention and its schedule and would lead to the cessation of the Convention and to unregulated whaling' (IWC 25, Chair's Report).

According to the Japanese commissioner, the proposed moratorium was:

> not only without scientific foundation, but in contradiction to the spirit of the convention, which stipulates that the amendments to the schedule of the convention shall take into consideration the interests of the consumer of whale products and the whale industry.

The US commissioner, however, said that his delegation would not have proposed the moratorium if it felt it were in contravention of the Convention. Argentina and France seconded the motion, which failed despite majority approval with eight votes in favour, five against and one abstention (IWC 25, Verbatim Record). In Day's words, 'when the crunch came and the moratorium vote was called for at the 1973 IWC, it was a very different story from the year before. This time it was the whalers who were reeling from the impact' (Day, 1992: 43–7). The four states which in 1972 had voted in favour of the moratorium were joined by Australia, Canada and France and, surprisingly, Panama – Denmark was the only state which abstained. Little changed quickly, however. In language rather prescient of debates to come, the Norwegian commissioner said of the moratorium that 'to try to press this matter through by a majority vote strikes me as being a somewhat unusual way in which to pursue international cooperation', and then suggested that the consequences of the passing of a moratorium through majority vote might well be the 'total collapse of the Commission as such' and that 'countries with no whaling interests at stake should have this in mind – thinking of the future of the Commission – when casting their votes' (IWC 25, Verbatim Record).

IWC 26, 1974, saw the Commission take an important step forward by adopting an Australian proposal and thereby introducing a new management procedure (NMP). This NMP:

> recognised that the management of whale stocks should be based not only on MSY concepts related to numbers and species but should include assessment of factors such as total weight of whales and interactions between species in the marine eco-system.

Again, a resolution, submitted by the US and seconded by Mexico, proposing a ten-year moratorium on all commercial whaling was considered. The Scientific Committee reviewed, and then considered still appropriate, its advice of the previous year. It 'drew attention', in addition, 'to the suggestion of possible competition between species whereby rebuilding of severely depleted stocks may not necessarily be maximised by a moratorium'. Competition between species is not, therefore, a new idea within the IWC. Australia, seconded by Denmark, then suggested an amendment which 'referred to the need to preserve and enhance whale stocks as a resource for future use and taking into consideration the interests of consumers of whale products and the whaling industry as required by the International Convention on Whaling' – language which Australia would undoubtedly not approve of today; but Australia was then still a whaling nation (IWC 26, Chair's Report).

The Australian amendment recognised that whale stock management should be based not only on the maximum sustainable yield (MSY) concept in number by species, but 'should also include such considerations as total weight of whales and interactions between species in the marine ecosystem'. The resolution was adopted by majority vote in the plenary session, although the words 'as a resource for future use' became 'as a resource for present and future use'. The US stated that it had voted for the resolution because it felt that it represented a significant step forward in the management of the world's whales, but that the US still supported a ten-year moratorium. A number of other commissioners expressed their agreement with this view (IWC 26, Chair's Report).

At IWC 27 and 28, in 1975 and 1976, the IWC wrestled with implementing the NMP, and New Zealand, presumably feeling that the IWC was heading in a direction it approved of, rejoined (Birnie, 1985). An important step was taken at IWC 28 with the Commission, for the first time, adopting a resolution calling for steps to be taken to limit expansion of the Alaskan bowhead fishery and to reduce the rate of whales struck and lost (ibid.). Both the Technical Committee and the

Scientific Committee pointed out to the Commission that steps should be taken to limit the taking, as well as the 'struck and lost' rate, of bowhead whales off North America, and that both the US and Mexico should be requested to counter 'harassment' of gray whales in breeding areas. Denmark proposed the adoption of resolutions to support both of the Scientific Committee's recommendations and, after amendment, both were adopted by the Commission (IWC 28, Chair's Report).

At IWC 29, in 1977, the US commissioner, in his opening address, delivered a message from US President Carter which included the suggestion that the US hoped 'to see all whaling nations, as well as interested non-whaling nations, join in the important work of the IWC so that we may create a truly effective and representative system to safeguard whales for the future' (IWC 29, Verbatim Record). The joining of the IWC by non-whaling nations, of course, has since become an important and contentious issue. Carter's words presaged the shift in membership in the late 1970s and early 1980s which was eventually to lead to the moratorium being adopted; but which was also to contribute to the present deadlock.

At IWC 29, in 1977, the Commission adopted its earliest resolution dealing specifically with the relationship between the ICRW/IWC and CITES:

> (1) The IWC offers to the CITES to act as its adviser on cetaceans. (2) The IWC accepts the CITES invitations to participate at the October meeting in Geneva. (3) The IWC will (a) advise the CITES of the conclusions of the [SC] regarding the status of whale stocks in relation to the CITES criteria; (b) provide advice at Geneva on any exploitation by non-IWC members of those stocks whose biological status is appropriate for Appendices I or II. (4) In addition, the IWC representatives should be prepared to discuss problems of identification of cetacean parts and products.
>
> (IWC 29, Chair's Report)

The Australian commissioner pointed to 'a potential problem with automatically absorbing the IWC's resolutions into CITES Appendices, as the two conventions meet two years apart and anomalies might therefore arise', and also pointed out that:

> the listing of a species in Appendix II of CITES has the connotation that species may become threatened without regulation of trade if trade is not regulated. In the case of whales, as of many other animals, this is not necessarily true; for example, in the case of whales which are not subject to harvest.
>
> (IWC 29, Verbatim Record)

As said, Australia was then still a whaling state.

The same meeting saw an accusation surface, which was eventually to become of significance, to become familiar and even today to be of crucial import. In Birnie's words, the:

> Japanese Commissioner accused some delegations of using the NMP as a means of achieving goals incompatible with it, specifically the moratorium; and warned that the international management of whale stocks would collapse if some states used their new 200-mile zones unilaterally to enforce their own ideas of management.
>
> (1985: 474–5)[5]

This does show something of the view the Japanese have always taken – that the oceans are a free commons. It also shows why the eventual 'Irish Proposal' of the mid-1990s – which included the premise that whales could be taken in, and *only* in, 200-nautical mile zones – was unlikely ever to gain favour with whaling states. Of course, in 1977 the Japanese concern was that they might be prevented from harvesting whales in other states' 200-mile zones.

A contentious issue which arose again was the aboriginal taking of bowhead whales. The Commission agreed to maintain the protected status of the stock and 'completely deleted' the exemption in the Schedule allowing aboriginal fisheries. This effective ban on bowhead whaling created problems for the US with regard to its obligations to protect the cultural rights and subsistence needs of the Eskimos; but the country undertook to 'implement the IWC regulation when US national legislation problems were removed, and meanwhile promised to undertake an intensive investigation of the size of the stock' (Birnie, 1985: 485–6). The 1977 ban on the taking of bowhead whales had a consequence which was to echo some 15 years later – in 2002, as will be seen – and then to re-echo.[6] After the US had lost the right to have its Eskimo people take bowhead whales, a Special Meeting was held in Tokyo in December 1977. At the Special Meeting the Commission was asked by the US to reconsider the earlier decision to prohibit the taking of bowhead whales from the Bering Sea stock and the Commission then agreed to a resolution which effectively overturned the earlier decision (IWC Special Meeting 1977, Chair's Report).

At IWC 30, 1978, the US put forward proposals for future action which included a catch limit set 'at the aboriginal subsistence needs' but which would not take more than 2 per cent of the stock, coupled with 'continued research'. After much debate and votes on different proposals, the Commission adopted (nine to one, with seven abstentions) a proposal that 'in 1979 hunting shall cease when either 27 have been struck or 18 landed'. At the end of the meeting the US asked for two more bowhead whales to be added to the 1978 catch limit, which was done (IWC 30, Chair's Report).

5 In 1976, the US had enacted a Fishery Conservation and Management Act which extended its fisheries jurisdiction to 200 nautical miles – eventually this distance would be reflected as 'codified customary international law' in the 1982 UN Convention on the Law of the Sea.

6 When, in May 2002, the US and Russia lost their aboriginal quotas, they immediately called for a Special Meeting, which was held in October of the same year. In 2002, as will be seen later, the US made concessions to pro-whaling countries in order to recover the lost quota, and it seems indisputable that similar negotiations behind closed doors preceded the Special Meeting in 1977. International diplomacy is seldom transparent.

The US also produced new figures for population estimates, arguing that its research programme had determined that the bowhead population was close to 2,300 whales, this being almost twice the size of the best estimate of the previous year on which the Commission had based its decision. The US argued that the 'risks imposed on the whale population by permitting a modest hunt are small when compared with the countervailing risks to human populations if the hunt is not permitted'. The US commissioner also said later that, at the time of his speaking, the Eskimo advisers to their delegation had returned home from London and had indicated that:

> in light of the vote yesterday, and what they assumed would happen today, they would not comply with what the IWC has done. I mention that because that means that we may have compliance problems of a significant nature. We may have a confrontation in the United States with native peoples. If the [lower] figure is accepted, I foresee non-compliance and blood on all of our hands and it may not only be the blood of Eskimos but the blood of more whales than we are talking about here today if the Eskimos do not comply.
>
> (IWC 30, Chair's Report)

The above is startlingly reminiscent of the arguments made in recent times, as to risk balancing and non-adherence, by the Japanese when pressing the claims of their coastal whaling people (whom the Japanese contend ought to be considered as being at least equivalent to aboriginal subsistence whalers).

As would eventually recur in 2002, other states took advantage of the US' difficulties. Birnie argues that it was to the advantage of whaling states to accede to the US' demands for an aboriginal quota as the concession 'undermined the US's previous efforts to press a moratorium on the IWC', and she contends that this problem 'persisted into future meetings and enabled the USSR, Japan and other commercial whaling states to bargain for higher quotas than might otherwise have been acceptable to the US and its supporters in the Commission' (Birnie, 1985: 500–1). The precedential value to pro-whaling countries of the US' awkward position in respect of aboriginal whaling is significant.

In 1977, in the Chair's Report of the IWC Special Meeting, it is recorded that the Secretary reported on the special working session of the CITES held in Geneva, which was attended by the Chairman and Secretary in October 1977; that cross-representation between the two conventions 'would be of mutual benefit, and the CITES was to offer Observer status to the IWC at meetings of the Parties'; and that the Commission 'agreed that the CITES should be granted observer status at its meetings and those of the Scientific Committee' (IWC Special Meeting 1977, Chair's Report). Cross-representation at observer level was in fact granted by each to the other. It was also recorded that the IWC Secretariat would provide direct assistance to the relevant UK authority which had offered to undertake a review of cetacean species for the Appendices to CITES (IWC 29, Chair's Report).

The Chair's Report for IWC 30 recorded that the CITES observer had made a statement 'welcoming the mutual granting of observer status agreed between the

CITES and the IWC', and, further, that CITES now had 46 member states 'including ten IWC members, and he appealed to those not in the CITES to join'. The Report concluded that 'there is growing cooperation between the two Conventions, and the IWC is invited to be represented at' the next CITES meeting in 1979 (IWC 30, Chair's Report).

Importantly, Peter Sand, the first CITES Secretary-General, suggested that the CITES Secretariat welcomed 'the establishment of working relations with the IWC and the mutual granting of observer status'. He suggested that there would be:

> several consequences which this growing cooperation between our Conventions will have. First, it will progressively eliminate the need for the provisional reservations expressed by some governments, vis-à-vis certain cetaceans listed on the appendices of CITES. These reservations were made before your Canberra meeting last year by four governments, Australia, Canada, South Africa and the USSR.

These reservations, per Sand, 'were limited to stocks of sei whales and fin whales and the reasons given in the case of Australia, for example, were precisely to preserve that country's position at the Canberra meeting of the [IWC]'. Sand then appealed to these states to withdraw their reservations 'in the light of the cooperative relationship established with the IWC', and appealed to those IWC member states which had not yet ratified CITES to do so. Interestingly, he suggested that one of the reasons why certain states – including Iceland, Japan, the Netherlands and New Zealand – might not have ratified CITES was because of concerns over the relationship between the two bodies. Sand then concluded with an appeal for a continuous and mutual exchange of information between the two conventions (IWC 29, Verbatim Record).

As in 1977, at the end of 1978 there was an IWC Special Meeting. The ostensible reasons for the holding of the Special Meeting were to set North Pacific sperm whale stocks for the 1979 season and to resolve certain financial issues. However, the parties took the opportunity to raise a number of other matters. One of these, a 'Resolution to the CITES', proposed by the US and seconded by the Netherlands, 'concerning support of IWC actions by the CITES' was 'adopted by five votes to four with six abstentions. Canada, Denmark, Iceland and Japan expressed their views on the difficulties of equating the IWC stock management categories with the CITES classifications and questioned what the Resolution might achieve' (IWC Special Meeting 1978, Chair's Report).

The resolution, as passed, read:

> Resolution to the CITES: WHEREAS, it is the purpose of the [IWC] to provide for the effective conservation and management of whale stocks, and WHEREAS, the [IWC] has adopted a New Management Procedure to carry out that purpose, and WHEREAS, the [IWC] has established regulations which allow no commercial taking of certain species and stocks of whales in given ocean areas, and WHEREAS, in order to contribute to the effort to

> conserve whales receiving commercial protection from the [IWC] and to reinforce adherence to [IWC] regulations, it is desirable to use each international opportunity to stop the taking and to ban trade in those species and stocks of whales which receive total protection from commercial whaling, BE IT HEREBY RESOLVED by the [IWC] that it request the Second Meeting of the Conference of Parties to the [CITES] to take all possible measures to support the [IWC] ban on commercial whaling for certain species and stocks of whales as provided in the Schedule to the [ICRW].

In debate on the resolution, the US considered it an:

> important opportunity to maintain working relations between two international organizations that have overlapping concern with citations (*sic*). The Commission now may take action to reinforce these relations and urge the parties to CITES to take action to ban trade in these species and stocks of whales protected from commercial whaling by this Commission.

Canada, however, argued that there was a problem with:

> trying to directly link the criteria used in the two items, one being the schedule of our convention and the other being the endangered species convention. And so, while there is a certain degree of complimentarity (*sic*) particularly with regard to the truly endangered species that we all recognize. This does not extend as directly to some of those species and stocks which are closer to the sustained management category.

This was supported by both Iceland and Japan. The US, however, argued that all that was being asked for was that CITES be asked to:

> take all possible measures, *possible* measures, to support the [IWC] ban on commercial whaling for certain species and stocks of whales as provided for in the Schedule. That is all that is asked for. There is nothing illegal about it.

The US contended further that there was 'nothing inconsistent with the international law', and that 'the proper parties to decide what animals of fauna and flora belong [in] what place in their Convention [are] the parties to that Convention. And presumably those parties will make that judgement under the criteria in that convention' (IWC Special Meeting 1978, Verbatim Record).

The Netherlands explained that in its opinion the resolution would 'leave sufficient room' to either Convention to take:

> whatever action they would take as a result of this request. This would not necessarily have to be the inclusion of the species in either one of the two appendices of the Convention but it could also be the adoption of the resolution to the effect of preserving a certain habitat or something.

Japan, however, argued that:

> if we are not requesting those species to be listed in Annexes 1, 2, 3, in certain cases we should make it specific to the Convention. Otherwise the resolution is too obscure, ambiguous, and it does give the impression that we are requesting those species to be listed in either of the three categories. So, we have to avoid in specific terms that implication. Otherwise I don't think this resolution carries any significance.
>
> (IWC Special Meeting 1978, Verbatim Record)

In the event, Argentina, Canada, France, Norway, South Africa and the UK abstained; Australia, Mexico, the Netherlands, New Zealand and the US voted in favour; and Denmark, Iceland, Japan and the USSR voted against. The resolution was therefore passed by majority vote.

Eventually, of course, parties were to do several 'flip-flops' and today sees the anti-whaling IWC member states arguing – broadly – that as long as there is a moratorium on commercial whaling in place under the IWC, CITES should not approve trade quotas for whale species, and that, generally, CITES ought to continue to recognise the IWC as the competent body for cetacean management decisions. On the other hand, Japan has turned to the argument that as the IWC is deadlocked – with anti-whaling member nations showing no willingness to end the moratorium on commercial whaling – CITES ought to step in and take trade-related management decisions.

By the mid-1970s, the US had begun to flex its muscles. The passing in its domestic law of the Pelly Amendment to the Fisherman's Protective Act gave the US the ability to take action where other states party to the ICRW did not, in effect, act according to the US' ideals. The US had therefore clearly decided to move toward direct action. However, it is important to recognise that many other considerations would play a role in the US' decisions on whether or not to use the weapons with which it had armed itself. Many vested interests come into play when the US chooses whether to act in international law or not. According to Day, for instance, the Packwood-Magnuson Amendment to the Fishery Conservation and Management Act needed to be passed by the US Senate, 'to force government's hand [in 1979], because of the American administration's refusal to use Pelly Amendment sanctions against Japan' (Day, 1992: 116). Domestic politics in the US, of course, means that nothing is simple and 'political will' is not always represented by the laws which happen to be on the books.

By the mid-1970s, it was obvious that whale populations were in serious trouble. It was estimated at this time that before whaling began there were approximately 3.9 million whales in the oceans of the world; but by 1975 whaling had reduced the number of whales to approximately 2.1 million, with the mature (therefore exploitable) population of whales having decreased from about 2.4 million to about 1.2 million. These figures, however, were described as being deceptive as: first, the selectivity of the whaling industry has caused proportionately much greater reductions in the populations of certain species, especially the larger baleen species, so that while worldwide whale populations had been reduced by about 46 per cent, some baleen whale species in the Antarctic had been reduced by as much as 96 per cent. Second, the figures on population reductions did not reflect reductions in whale biomass; the point being that selective depletion of blue and fin whales had caused an 85 per cent reduction in the total baleen whale biomass in the Antarctic, from an estimated 43 million metric tonnes to 6.6 million metric tonnes (Scarff, 1977). Proponents of the ecosystem approach in the mid-1970s could argue that whales generally were in trouble; today, however, the same argument – selective depletion – is used by proponents of a return to commercial whaling, as will be seen.

By IWC 31, in 1979, membership had increased to 23 – the new members, however, being more pro-use than pro-conservation (in the 'preservation' sense). Of importance, though, was the joining of the Seychelles – a tiny country which was to punch above its weight and which had an immediate and important impact, arguing in favour of creating whale sanctuaries. The bowhead whale issue arose again with the US seeking an aboriginal quota even though the Scientific Committee had advised that no quota be allowed. At the same meeting an initiative by Panama to eliminate sperm whaling in the Antarctic by placing a moratorium on whaling by factory ships in that area was accepted. Nevertheless, it was accepted only with major compromises – an exclusion of minke whales being agreed to accommodate Denmark, and the area of exclusion being reduced to accommodate both Japan and the USSR (Birnie, 1985). One point to note is that the moratorium was becoming the most popular solution proposed – probably, its non-permanent nature made it seem inherently a compromise. The idea of compromise was becoming entrenched – not altruistically, however, but in self-interest.

At IWC 27, 1975, the Commission had discussed catch quotas in areas where non-members took significant numbers of whales and agreed to invite these whaling nations to join as their activities were thought to be weakening IWC's systems. The UNEP observer to the IWC had commented on this issue, saying that his organisation had already approached non-member whaling nations, 'but had received negative replies'. The Latin American states, it seems, had their own organisation and were therefore reluctant to commit to the IWC. A small working group was then set up, comprising Canada, Denmark, Iceland, the UK and the US to review matters. The Commission also adopted a resolution which, inter alia, 'requested that the non-member states cooperate to avoid over-exploitation and join the Commission' (IWC 27, Chair's Report). This is an interesting thing as, although it was active whaling states which were targeted, it makes it clear that the IWC has never been averse to bringing new states on board; in fact, that it has been active policy. In due course, and into the future, this was to change, and may yet change, the face and nature of the IWC.

At IWC 29, 1977, a US-proposed, Canadian-seconded, and then Canadian-amended, resolution was adopted on the Prevention of Importation of Whale Products (IWC 29, Chair's Report). Japan objected to the resolution because, in the words of its commissioner, 'my delegation cannot find it possible to associate with this resolution, because of the implications of GATT. Also, I think it is not good practice to penalise non-member countries by imposing these kinds of regulations' (IWC 29, Verbatim Record). At this stage, then, Japan was opposed to the involvement of other bodies than the IWC in whale management and trade decisions.

The year 1979 has been described as important in the history of whale conservation, with 'conservationism' having become a 'very powerful force at the IWC'. M'Gonigle suggested that although the negotiations of that year 'produced only another incremental decline in the quota level, the process of achieving that reduction and the nature of the quota itself represent a major change from the past' (1980: 170–1). Six new members, which brought the total membership to 23, included countries such as the Seychelles and Sweden, which had joined specifically 'to advance the cause of whale conservation'. Although the four other new members – Chile, Korea, Peru and Spain – were whaling countries, the balance was shifting away from pro-whaling countries as old members – Australia, New Zealand, the UK and the US – changed their policies (M'Gonigle, 1980).

The issue of a possible worldwide ban on whaling was discussed in 1980, and the IWC's Technical Committee, by a majority vote, even recommended a French- and US-proposed amendment which would implement a moratorium on commercial whaling. Voted on, the proposal did not achieve the three-quarters majority necessary (the vote went 13 to nine, with two abstentions) (IWC 32, Chair's Report).

The moratorium on commercial whaling was finally adopted in 1982 – ten years after being first proposed – and only after a number of developing countries had joined, which Birnie considers significant (1985b: 952–3). Obviously, it was not solely the developing countries that pushed the moratorium – in fact, it was a number of important developed countries that switched their positions. And, too, it seems likely that most of the developing countries were strongly influenced by developed countries, or by powerful non-governmental organisations (NGOs) within developed countries.

Attempts, pushed especially by the US, were made in the late 1970s to hold a new conference outside of the auspices of the IWC/ICRW. These attempts, however, foundered on the rocks of the jealous guarding of sovereign rights, and of the impasse that loomed immediately between states which wished to use their Exclusive Economic Zones (EEZs) to prevent any whaling at all, and those which wished to use the opportunity provided by a new convention to reduce, rather than increase, the powers of the IWC. It is interesting to note that the countries which appeared most to want to reduce the authority of the IWC were Brazil, Canada and South Africa – Canada produced a draft for a new convention; Brazil and South Africa suggested amended texts. Eventually the Secretariat, working with these three texts, produced a composite, and Revision Conferences were attempted in 1978 (Copenhagen), 1979 (Estoril) and 1981 (Reykjavik). Ultimately, though, no revision proved acceptable and the parties did not even come close to agreeing; with fundamental disagreement even as to the purpose of the ICRW (Birnie, 1985b: 955–6).

Birnie does comment, though, that IWC 31 was extremely significant. In particular, she refers to the involvement of the Seychelles – which seems to have taken the lead at a critical juncture (Birnie, 1985b: 959–60). However, the entry of this small developing state was followed by more – as states realised that they needed a way of breaking the deadlock on the IWC, so they began to lure in new members sympathetic to their causes. At this time, the states joining appeared generally to be on the 'conservation' (anti-whaling) side. M'Gonigle writes that four overlapping blocs became 'entrenched'. These blocs were 'the nine whaling nations';[7] 'coastal states, led by Canada, which sought to restrict IWC jurisdiction within their 200-mile exclusive economic zones';[8] the US, 'which demanded a large bowhead quota for the Alaskan Eskimos in spite of the committee's zero quota recommendations'; and the 11 countries which took a 'conservationist' stance.[9] As a result of compromises, in M'Gonigle's view, the IWC rejected the moratorium proposal and many of the Scientific Committee's recommendations for lower or zero quotas and granted the US a substantial bowhead quota (1980: 237).

At IWC 32, in 1980, membership grew to 24. Oman became the first Middle Eastern state, and Switzerland the first landlocked state, to join. With all members attending, there were, basically, nine whaling and 15 non-whaling states. Interestingly, Taiwan – which was at the time a whaling state – had also applied for membership but its application had been rejected by the depositary government (the US) on the ground that only two of the IWC's members – Korea and South Africa – then recognised Taiwan's government (Birnie, 1985: 600–1). Today, Taiwan remains a major fishing country outside of both the IWC and CITES.

IWC 32 saw the impasse continue. After this meeting, however, the so-called 'conservationist' (anti-whaling) states increased their efforts to attract new members. Not all of the new members were particularly committed to, or even very interested in, the IWC. Dominica, Egypt and Jamaica, for instance, apparently made only token efforts at participation. Canada decided to leave the IWC altogether – not wishing, probably, to face the same internal division as the US over the aboriginal whaling issue (Birnie, 1985b: 962–3). It was probably in deep frustration at the IWC that Germany was to argue in the same year, at a CITES Technical Meeting, that 'the old whaling convention is no longer suitable' and suggest that CITES could 'go beyond' the IWC's standards (Osterwoldt, 1982: 70).

The expanded membership assisted the 'conservationist' states almost immediately. At IWC 33, 1981, the moratorium to come was clearly presaged as an indefinite moratorium to be placed on sperm whale hunting – the vote went 25 to one (Japan voting against), with three abstentions (China, Iceland and the USSR).

Also at IWC 33, many of the new members made opening statements which clearly indicated their support for the move toward a moratorium. St Lucia's

7 Brazil, Chile, Iceland, Japan, Korea, Norway, Peru, Spain and the USSR.

8 Argentina, Brazil, Chile, Mexico, Peru and Spain supported Canada's arguments.

9 Argentina, Australia, France, Mexico, the Netherlands, New Zealand, Oman, the Seychelles, Sweden, Switzerland and the UK.

commissioner, for example, suggested that 'oceans and their flora, fauna and mineral resources' were 'humankind's common inheritance'; although countries such as Costa Rica, India, St Vincent and Uruguay indicated that, while they agreed with this 'common heritage' approach, they were not in principle opposed to sustainable utilisation. St Vincent suggested that whales, if their population levels recovered, could be used again. India gave an interesting indication of their views in supporting 'the concept of conservation coupled with sustained (*sic*) development', and in emphasising support for CITES (Birnie, 1985b: 962–3). Birnie comments that most developing state members supported a global moratorium, at least on a ten-year basis, but that many still wanted to keep the option 'of exploiting whales not yet regulated and also to leave open the long-term possibility of resuming whaling if stocks ever recover' (ibid.: 963–4). This argument does, of course, mean that countries such as St Lucia are not being hypocritical when they argue as they sometimes do fervently, and vote in the twenty-first century, in favour of sustainable commercial whaling.

Membership increased to 32, with the eight new members[10] being mostly states favouring 'conservation' rather than exploitation of whales. In its (verbal) opening statement, India drew attention to the symbolic value of the whale – suggesting that India 'shared the concern for preserving for future generations the inestimable natural resources represented by the cetaceans which is in itself a symbol of the much broader concern for the preservation of man's environment' (IWC 33, Verbatim Record). India then sought to link the IWC/ICRW to CITES, arguing that:

> the objectives set forth in the International Convention of 1946 are far from being realised. There is world-wide concern on this score, amply reflected in a resolution passed by the Third Meeting of the Conference of the Parties to the Convention on International Trade in Endangered Species of Wild Fauna and Flora, held at New Delhi four months ago, whereby the sperm whale, the fin whale and the sei whale have been placed on Appendix I of that Convention. As Chairman of the Standing Committee of CITES, representing the interests of all the Contracting Parties, which now number seventy, India considers it necessary to draw the attention of this gathering pointedly to this important development and to urge the need for a similar consensus in our deliberations here.
>
> (IWC 33, Verbatim Record)

Jamaica, in its (verbal) opening statement, stated categorically that 'Jamaica has always felt a sense of shock at the very thought of the slaughter of a mammal of the proportions of the whale' (IWC 33, Verbatim Record). Given the ferocity with which this opening statement was delivered, it is of interest to note that Jamaica is

10 China, Costa Rica, Dominica, India, Jamaica, St Lucia, St Vincent and Uruguay.

not, in 2013, an IWC member. Other Caribbean countries, such as St Lucia and St Kitts and Nevis, tend strongly, in the twenty-first century, to be supporters of sustainable use – so much so that many Western NGOs have accused them of having been bribed by Japan. Unlike Jamaica, another Caribbean nation, St Vincent and the Grenadines, was, and remains, a whaling nation – albeit aboriginal subsistence whaling, and in small numbers. In its (written) opening statement, the country suggested that it would like to see whales become 'useful to them' again – if whales should recover 'as a result of conservation actions by the IWC and by protective measures taken by states within their zones of national jurisdiction' (IWC 33, Opening Statements).

Of great interest for the present book is the fact that, although the country was only an observer at the meeting, Tanzania submitted a (written) opening statement, stating that it:

> wishes to express solidarity with IWC in its endeavour to ensure rational exploitation of whales which are threatened with extinction because of their indiscriminate and cruel exploitation. Tanzania being a custodian of one of the largest heritage of wildlife cannot fail to realise the danger posed to marine life by disorderly management of whales. We deplore the rapid killings of whales which has manifested itself in all oceans.
>
> (IWC 33, Opening Statements)

Tanzania is today generally a supporter of Japan's views in the IWC, and it is interesting to see the country drawing links between whales and its own terrestrial wildlife at so early a meeting. This was of course a theme of the time. Along the same lines, Oman, in its (written) opening statement stated that it believed it to be the 'duty of all member states, whaling or non-whaling, to conserve these resources for the future generations' and explained that 'we have learned from the mistakes of other people whose activities led to the complete extinction of some animals such as Dodo of Madagascar (*sic*)' (IWC 33, Opening Statements).

This meeting was apparently conducted in a much less tense atmosphere than others of recent years – possibly because of the settlement of the Alaskan bowhead quota on a three-year-block basis. However, the meeting was not without important developments. Birnie's view is that a 'quiet revolution occurred' – with pressure increasing for various prohibitions on whaling to be adopted. In the end, one of these proposals – for an indefinite moratorium on sperm whaling – was adopted by a large majority (Birnie, 1985: 609–11). The moratorium, clearly, had become the strategy of choice for the 'conservationist' states – its non-permanent nature making it more attractive to those states reluctant to give up the right to whale.

An interesting description of the influx of new members can be seen in the (written) opening statement to IWC 33 by the USSR, which stated that:

> there is an increase in the number of countries which now have invaded the IWC having never had any relation either to exploration or to utilization of

> whale resources. We believe that this process in the Commission is abnormal, and we ought to seriously consider its consequences.
>
> (IWC 33, Opening Statements)

In the political context of the early 1980s it seems unlikely that the word 'invaded' can have been misunderstood! The opposing view can, however, be seen in the words of the commissioner for St Lucia at the same meeting: that with the entry of new members 'this Commission has in my view taken a giant step forward in terms of becoming what it should be' (IWC 33, Opening Statements).

In its (written) opening statement, Japan showed clearly that it was worried by so many new members joining. It stated that:

> we must stress the rights of the contracting governments, as spelled out in the preamble and relevant provisions of the Convention, cannot be denied or challenged merely by a vote at the Commission, which is empowered solely to implement the Convention

and that 'should such a proposal be carried by virtue of the number of the votes, we would reserve every legitimate right available to us to uphold our cause' (IWC 33, Opening Statements). At the time, then, Japan was not in favour of allowing the membership of states not directly interested; but, as with so many things in the whaling arena, this attitude would change.

3.3 The moratorium

IWC 34, 1982, turned out to be one of the most significant meetings ever. By now there were 39 members and 37 attended.[11] The increased membership gave the 'conservation-minded' nations sufficient numbers to achieve a successful moratorium vote – 75 per cent being necessary. It was the Seychelles which made the proposal that was adopted, this being for a moratorium achieved by way of a zero quota on all commercially exploited stocks. The term 'commercial' was not, however, defined. After much debate and several amendments, the proposal was voted on. The vote saw 25 in favour, seven against, with five abstentions, and was therefore passed.[12] The amendment to the Schedule, as finally adopted, read:

> [n]otwithstanding the other provisions of paragraph 10, catch limits for the killing for commercial purposes of whales from all stocks for the 1986 coastal

11 The eight new members at IWC 34 (Antigua and Barbuda, Belize, Egypt, Germany, Kenya, Monaco, Philippines and Senegal) were all non-whaling states.

12 The vote on the moratorium (the zero quota) in 1982 was as follows: in favour, Antigua and Barbuda, Argentina, Australia, Belize, Costa Rica, Denmark, Egypt, France, Germany, India, Kenya, Mexico, Monaco, the Netherlands, New Zealand, Oman, St Lucia, St Vincent, Senegal, the Seychelles, Spain, Sweden, the UK, Uruguay and the US; against, Brazil, Iceland, Japan, Korea, Norway, Peru and the USSR; abstentions, Chile, China, Philippines, South Africa and Switzerland.

> and the 1985/86 pelagic seasons and thereafter shall be zero. This provision will be kept under review, based upon the best scientific advice, and by 1990 at the latest the Commission will undertake a comprehensive assessment of the effects of this decision on whale stocks and consider modification of this provision and the establishment of other catch limits.
>
> (Schedule, para. 10(e))

The moratorium was therefore made dependent on an undertaking to review the move, after comprehensive assessment, by 1990. Peter Best told me that this had been an inclusion (in his words, a 'sop') designed specifically to bring Spain aboard (pers. comm., 2006). At the meeting, Spain made the point that it did not consider the moratorium to be a 'total ban' but rather a 'temporary interruption of the activity which does not from the part of Spain result from other considerations than those of a need to add very carefully in managing these very important stocks of marine living resources', which does tally with Best's contention (IWC 34, Verbatim Record). Douglas Butterworth confirmed this, telling me that the moratorium 'was essentially a deal with Spain'. It was, he said, 'really, like all these things, a compromise – the review within ten years, it was all part of the plot because it was a question of one vote here or there whether it went through'. If it was merely a ruse, it was one which may yet have serious consequences – according to Butterworth it was a commitment which 'countries like Japan took literally' (pers. comm., 2006).

Something of the heated nature of the matter can perhaps be seen in the recorded words of the (Argentinian) Chair immediately after the vote was taken: 'Silence, silence please. I think it is not the behaviour of the observers and the representatives here to express the situation in that way. Thank you' (IWC 34, Verbatim Record).

Japan argued against the moratorium proposal on the basis that the Scientific Committee had 'ruled many times in the past that there is no scientific justification or the biological need for a total or blanket moratorium' and that the moratorium was a political issue; whereas the Convention provides that any regulations should be adopted 'on the basis of a scientific justification'. The Japanese commissioner then argued that the 'proposal is clearly in violation of the present Convention' and that it 'should be beyond doubt' that the rights of contracting governments, in the ICRW and in other relevant international law, 'cannot be denied or advocated (*sic*) on the basis of a vote of the IWC'. The Norwegian commissioner argued that he considered the moratorium 'incompatible' with the Convention; the proposal, he suggested, 'distinguished between one type of whaling which comprises several other types of whaling lumped together under the term commercial whaling and other types of whaling'. Iceland argued that 'adoption of a moratorium on all commercial whaling based on strength of votes but not on scientific findings' was 'most certainly contrary to the fundamental objectives and purposes of the Commission'. Korea agreed, arguing that 'any proposal which is short of scientific data defiles the basis (*sic*) objective of the IWC' (IWC 34, Verbatim Record).

Interesting contributions to the debate, particularly given the future stances of parties, came from countries such as Antigua and Barbuda which argued that 'there is a greater consideration than the numbers game my colleagues from Japan would have us believe is truly the issue'; which greater consideration was that there is 'no humane method of continuing this needless form of industry'. St Lucia argued that at the time the Convention was adopted, in 1946, whale stocks were not as depleted as at present and 'therefore there were no provisions made at the time for the circumstances with which we are now confronted' (IWC 34, Verbatim Record). Ironies abound. Today, these countries appear to be firm backers of Japan's efforts to resume commercial whaling. Brazil, on the other hand, voted against the moratorium, but is now a firm opponent of resuming commercial whaling.

Japan, at the end of the meeting, made a statement in which it expressed its anger over the moratorium's adoption. 'The global cessation', according to the statement,

> of commercial whaling, in itself incompatible with the objectives and purposes of the ICRW, was proposed without any scientific basis and adopted through the sheer power of votes. This is totally unacceptable to those who share serious concern about the future of the IWC.
>
> (IWC 34, Plenary Report)

On many subsequent occasions, Japan has questioned the future of the IWC in a similar manner.

On the relationship between the IWC and CITES, questions were raised from the Scientific Committee as to how best to advise CITES on cetacean matters. The Scientific Committee recommended that it proceed by forwarding to CITES relevant information from its reports. The IWC approved this procedure. Some members of the Scientific Committee (supported by Denmark), however, argued that the 'questions being asked and the criteria being used by CITES for the listing of organisms' were quite different from those being used by the IWC in classifying whale stocks, and that, consequently, if meaningful advice were to be given to CITES the Scientific Committee should provide specific advice in respect of the appropriateness of listing certain species on CITES Appendices. On CITES, also, St Lucia argued that, while it had sympathy for countries engaged in whaling, the IWC represented 'a very considerable strength of world opinion'. St Lucia suggested that not only the nations present who supported the moratorium proposal, but also 'many other nations who have joined CITES proclaim that several species of whales should be completely protected'. This, according to St Lucia, 'clearly represents the majority of opinion in the world' and 'the nations who currently whale should consider why it is that so many nations who have nothing to gain should want to stop whaling' (IWC 34, Chair's Report). It is ironic that the worm has so turned that in recent times, in the twenty-first century, the argument is made by countries in favour of commercial whaling (including St Lucia) that it is now CITES which better represents the great variety of world opinion and that this opinion is largely in favour of commercial use.

Birnie comments that the impact of developing countries in the IWC was dramatic and that their participation 'changed both the nature of the Commission and its mission' (Birnie, 1985b: 965). Nowadays, the charge is often made that developing countries are 'pawns' – but this is to disparage the vital role of pawns. Developing countries may yet play important roles in determining the ultimate fate of the ICRW.

That many of the contracting governments argued that the moratorium was not based on scientific evidence provokes an important comparison with elephants – neither 'ban' being based scientifically on 'endangered status'.[13]

To a large extent, of course, the moratorium had its basis in powerful national demand that whaling cease. It would be interesting, and instructive, to see whether a boycott call would be effective today. The world may well have changed its views and consumers might be less concerned than in the early 1970s with the protection of whales generally. There is a strong argument to be made that it was the consumer boycott of ivory products in Western countries that truly brought the ivory trade to a standstill, at least until recently, and not the CITES Appendix I listing; but it may be that there is today a new – and arguably more sophisticated – understanding of the issue.

Komatsu and Misaki write that IWC 34 saw a dramatic increase in IWC membership, and that:

> during this decade, whaling nations such as Korea were recruited to join the IWC. They joined on the advice of the world's Superpower, but were stuck with a reduction of quota, sometimes to zero depending upon the stock. There were also new recruits from the small island nations of the Caribbean, which had their annual fees paid by anti-whaling organizations.
>
> (2003: 89–90)

This charge is, of course, extraordinarily ironic given that Japan had already been charged with exactly this conduct – having allegedly brought Brazil in in the 1974/5 whaling season to vote in line with Japan – and given that Japan is today accused regularly of bringing in small nations to vote as it dictates. In irony upon irony, in fact, it is now to a large extent small Caribbean states which support Japan. Where the 'blame' lies for this tactic is probably impossible to determine. It might even be impossible to determine whether it was 'pro-whaling' or 'anti-whaling' states which first adopted the tactic. Of course, this probably does not matter. It seems that at different times both sides have resorted to the tactic in order to win support for their arguments – and crucial votes. Hypocrisy is not the preserve of single protagonists in international environmental law and negotiation. What else does this tell us about international law and negotiation? Perhaps that it is – at least in good part – about learning ways to manipulate.

13 This is something that makes comparison between whales and elephants intriguing. Neither African elephants nor minke whales are numerically on the point of extinction, and this allows proponents of 'sustainable use' to argue that the species are being protected for reasons other than scientific.

By the late 1970s, the tide of public opinion had turned – perhaps irrevocably – within many former whaling nations. Pressure was brought to bear upon a number of nations still involved in whaling, but not all were eager to give up whaling, or even pirate whaling (Day, 1992). Whaling ban proponents likewise found it difficult, sometimes, even to persuade those countries ostensibly against whaling to act. Rose and Paleokrassis write of 'increasing reluctance' by the US to:

> use or threaten measures against nations who breach IWC regulations. What has tended to occur is that sanctions are threatened but subsequent negotiations result either in reduced action or none at all. Similarly, on the occasions when sanctions have been applied, their effect has often been negligible.
> (1996: 170)

They suggest that despite successes achieved by the US in 'promoting adherence to and compliance with the ICRW through threatening economic sanctions', the country is wary of taking action which might damage its national interests, and conclude therefrom that there are 'limitations to what an individual state can do or is willing to try to achieve for the promotion of a common good' (ibid.). These suggestions are significant. States have (sometimes conflicting) domestic constituencies they must satisfy, and while trying always to act to their overall best advantage, there is little room for altruism. US reluctance to follow through with its threats meant, per Vogler, that the 'scientific whaling loophole' was to be exploited 'on a huge scale by the USSR and to a lesser extent by the Japanese' (Vogler, 2000: 53). To a large extent, then, the IWC remained an 'old boys' club' through the 1970s and even into the mid-1980s.

As early as 1978, accusations were made of undue Japanese pressure on other members. M'Gonigle claims that in 1978 the moratorium had been put on the agenda by Panama, but before the meeting began Panama dismissed its commissioner and then withdrew the proposal amid allegations that the Japanese government had applied economic pressure to Panama 'by threatening to cancel a ten million dollar sugar purchase' (1980: 160–1). While the truth of this cannot be known, accusations that this sort of pressure is a tactic used by Japan have been made at least from the late 1970s until the present day. Of course, the forms of pressure put on 'allies' by the anti-whaling countries might be little more morally righteous.

It does appear that in the early 1980s the whaling nations were on the back foot. Birnie's general assessment is that there was more pressure put on whaling states as:

> more states adopted stricter national laws, more participated in the new international conservation treaties, more (especially non-whaling states) joined the IWC, the EEC adopted binding regional measures and NGOs activated an intensive campaign to ban whaling for the time being.

All of this led, she suggests, by 'incremental adoption of more conservatory measures' to the IWC moving 'closer towards protecting depleted whale populations rather than sustaining the industry at the expense of the whales' (Birnie, 1985: 575).

Technically, it is not correct to use the word 'moratorium' to refer to the provision of a 'zero quota on catch limits' in paragraph 10(e) of the Schedule to the ICRW. While 'moratorium' is clearly more correct than the word 'ban' – as often popularly used – it is not as accurate as would be 'zero quota'. Nevertheless, the word 'moratorium' has been so used by IWC members themselves that it is certainly the most accepted term today.

IWC 35, 1983, saw two new members join: Finland and Mauritius. Finland, in its opening statement, said that it supported the 'coming moratorium on commercial whaling' – but that it did regard whales as living resources and as 'shared international resources' requiring international cooperation for their conservation (Birnie, 1985: 624–5). According to Esko Jaakola, former long-serving IWC commissioner for Finland, Finland joined originally in order to support the moratorium – telling me, perhaps a little tongue-in-cheek, that it had seemed to him that the Greenpeace representative would 'not go away' until Finland joined (pers. comm., 2007).

The next battle became to persuade those states that had lodged objections to the moratorium to withdraw these objections. Japan, Norway, Peru and the USSR had lodged objections; other whaling states – Brazil, Chile, Iceland, Korea and Spain – had not, possibly because of the IWC's commitment to review by 1990. The US was putting pressure on Norway, in particular, (under the Pelly Amendment, which would restrict Norwegian fishing industry exports to the US) because of Norway's objection to the zero quota and its objection to abandoning using cold harpoons, and Japan. The US targeted Japan under the Packwood-Magnuson Amendment, threatening to prevent Japan from catching the quota of fish previously allocated to it by agreement in the US' 200-nautical-mile zone (Birnie, 1985).

IWC 35 saw a new argument put forward by both Japan and Norway. This was to the effect that both countries had coastal communities in remote regions, to whom whaling was both a vital economic interest and so rooted in history and culture that it was arguably a new form of, or something akin to, subsistence whaling – although distinct from the aboriginal subsistence whaling already permitted under the ICRW (in terms of the Schedule). This is an argument that is still being made in the twenty-first century, and which has not yet been accepted by anti-whaling IWC members.

The US was putting pressure on members it saw as recalcitrant; but, at the same time, all was politics and the US also showed a reluctance to follow through with its threats, such reluctance verging even on hypocrisy. This charge comes through in Day's words, when he tells us that 'Japan had for years used the IWC as a defence against the United Nations' call for a moratorium'; but that 'when in 1982 the IWC also called for an end to whaling by 1986, Japan switched tack again', saying it 'saw no legal or moral obligation to accept any decision of the commission on such a proposed moratorium'. This defiance would not have been possible, alleges Day, 'in the face of the drastic sanctions of the Packwood-Magnusson Amendment, but the American administration once more betrayed its avowedly conservationist stand' (1992: 157). The effect of this was that all of Iceland, Japan, Korea and Norway continued to hunt whales – the latter two countries through the 'scientific research' exemption, and Norway under its objection. 'Bizarrely enough', comments Day,

> for the first year of the moratorium zero quota in 1985–86, the whaling nations killed more whales (7 217) than was permitted the year before (6 623). In 1986–87, another 6 361 were killed under objection or scientific permit, despite the official position of a zero quota.
>
> (ibid.: 158–60)[14]

Day ascribes this 'defiance' largely to Japanese economic power. Certainly, Japan's economic might makes it a formidable opponent, even as this very economic might is what makes it so important a protagonist in the environmental debate.

3.4 The philosophical battle

The philosophical battle was not won in the 1980s, as overwhelming as sometimes seemed to be the support for the preservation of the whale, and its 'iconisation' as a symbol. Vogler writes that the prohibitionist view 'seemed in the mid-1980s to have majority support in the IWC, but was bitterly contested by the few remaining whaling nations, notably Norway, Japan and Iceland'. He adds that no less a person than Gro Harlem Brundtland 'joined the contest in support of a resumption of catches by the whaling communities of northern Norway', and quotes her as saying that we must:

> base resource management on science and knowledge, not on myths that some specifically designated animals are different and should not be hunted regardless of the ecological justification for doing so. International cooperation is in danger if this kind of selective animal welfare consideration is allowed to dictate resource policies.
>
> (Vogler, 2000: 51)

The ICRW is open for all states to join. de Klemm and Shine suggest that this is why the IWC 'has gradually been transformed from a commercial exploitation treaty into a conservation treaty' (1993: 48). The assertion that the ICRW *has* been so transformed into a preservationist (as they mean by 'conservation') is debatable, as will be seen.

14 These and similar figures may be a crucial element in Australia's current litigation against Japan in the ICJ, in which it argues that Japan's research whaling programmes are in violation of the ICRW.

According to Andresen, the present trend, generally welcomed, is toward 'increased participation by states, as well as nonstate actors, in international regimes'; but he argues that within the IWC, NGOs are so well represented in the delegations of key nations that they exercise a disproportionate influence, thereby representing 'a rare case in which NGOs, usually conceived of as the "underdogs", are allied with the most powerful players'. His conclusion is that it is arguable that this NGO influence has been less than constructive as they may have turned the issue 'into a simple battle of whaling versus non-whaling' (Andresen, 2001: 240–1, 241). This is an important point, but Andresen may have over-simplified it. The issue is often framed as though it has become a 'simple battle', but it is far more complex than this – and much is at stake. Also, it is worth noting that many NGO representatives have attended the IWC for years, even decades, while commissioners change frequently. In many ways, the NGOs now provide a crucial repository of institutional wisdom.

Andresen then makes an intriguing claim. He argues that the time may have come for 'environmentalists to realize that they have won the battle to save the whales'. The whaling issue, he suggests, was:

> an excellent first target for the young international environmental movement in the early 1970s; the whaling industry had diminished greatly and was therefore very weak, the problem was easy to visualize, and massive failure to conserve the largest animals on earth could be easily demonstrated.

In the 1990s, however, 'when *sustainability* is the key word among environmentalists', he concludes that 'in *rational* terms it is difficult to understand the massive opposition toward carefully managed harvesting of species that are clearly *not* threatened' (Andresen, 2001: 249–50).

Gambell supports this, arguing that anti-whaling groups have been forced of late to shift their justification for being opposed to whaling, with more emphasis being placed 'on the methods of killing whales and the length of time that elapsed before a whale died from harpooning' and with the 'ethics of the pain and suffering inflicted' becoming an issue (2003: 76–7). Iino and Goodman claim that the 'Save the Whale' slogan continues to raise hundreds of millions of dollars for anti-whaling organisations, but is now 'irrelevant' as 'whales have been saved' (2003: 11). Certainly, it might be true that framing the debate as a simple fight to save endangered whales is disingenuous, or at least an over-simplification – and perhaps even owes more than it should to the ease of using the whale as a symbol to raise money – but what this gives too little credit to is the possibility that it is important to the environmental cause to retain the whale *as* a symbol. After all, it is the pro-whaling camp which appears nowadays to have cornered the market on the 'ecosystem' approach to conservation in its arguments for the resumption of whaling. Put another way, the battleground may have shifted somewhat – but the whale remains an important symbol, like the banner to which the legions of an army rally wherever the battle might shift to on the field.

When Canada withdrew from the IWC in 1981, it gave as its reason for doing so the argument that the ICRW was intended to 'provide for the proper conservation of whale stocks and thus make possible the orderly development of the whaling industry', and that since the Canadian government had, in 1972, banned commercial whaling it no longer had 'any direct interest in the whaling industry or in the related activities of the IWC'. This clearly implies that Canada did not believe that the purpose of the ICRW was separable from commercial whaling (IWC 33, Opening Statements). Many commentators agree. Jacobson argues, for instance, that 'an evolutionary change in custom favouring protection of whales from whalers, if such a change is indeed on the horizon, should not as easily be seen as automatically amending the treaty'. His argument is that the 'Vienna Convention's strong emphasis on the ordinary meaning of a treaty's language is not helpful to those who wish to qualify the ICRW as a whale preservation agreement', and that 'analysis of the whaling convention's language setting forth its purposes and objectives, especially when viewed in light of its context, reveals that it is among the clearest wording of any international agreement'. Scornfully, he suggests that 'it has required some linguistic gymnastics from those analysts who have attempted to manipulate this wording to justify the IWC's actions in purporting to establish a Southern Ocean sanctuary and an indefinite moratorium on commercial whaling'. In particular, he explains that such verbal gymnasts have focused on the word 'conservation' (in the sense of 'preservation', obviously) in the wording of the Convention; but simultaneously have ignored the 'other principal part of the formula that very clearly establishes whale harvests as the ultimate objective, as if the IWC has been given an option to choose between them'. He concludes that the ICRW's 'frequent references to its purposes and objectives, taken collectively, easily support a plain-meaning reading that identifies the conservation goal as a *means* of achieving the more important sustainable-whaling *end*' (Jacobson, 2001: 86, 87–8). Jacobson does not appear to allow, however, for the possibility that a purpose might indeed change.

Lyster argues that change is possible; but cautions that this is allowed only when 'the ordinary meaning of the text is ambiguous or obscure' or when such ordinary meaning leads to a result 'which is manifestly absurd or unreasonable'. He concludes that 'an interpretation which fulfils, rather than frustrates, the objectives of the treaty should be preferred if there is any ambiguity' (Lyster, 1985: 7–8). Paradoxically, a little *more* ambiguity in the ICRW might not have been a bad thing if it had led to more flexibility. That the treaty is definite in its terms has arguably contributed to the current polarisation within its membership – there is little room for compromise where only one of two competing interpretations can be correct. Of course, one must remember that Birnie has argued that it *is* a flexible treaty. Arguably, however, this is because she was writing at a time when – the 1982 moratorium having just come into force in 1985/6 – it appeared that the treaty had, in effect, been adapted. In more recent years, the impasse between camps in the IWC appears set in stone.

Iino and Goodman describe the ICRW as having been one of the first resource management conventions to embody the principle of sustainable use, but they contend that the Convention has become:

> dysfunctional because the majority of its members no longer support the purpose of the Convention, its decisions are made on the basis of emotional and moral judgments rather than scientific evidence, and it is becoming increasingly irrelevant since most whaling, even that conducted by IWC members, occurs outside of its control.
>
> (2003: 4–6)

This last contention, for which the writers offer no evidence, seems highly unlikely – unless it is the annual takes of small cetaceans by Japan that is being referred to.

Vogler notes that internationally, attempts to bring open access fisheries under some form of management regime have long posed problems (2000: 48–9). This simply reflects the generalised difficulty of dealing with a resource open to exploitation by all – but the whaling situation is further complicated by its history, symbolism, emotion and by both altruism and self-interest.

Friedheim praises the text of the ICRW, going so far as to describe it as being, despite its age and the fact that it is one of the oldest international environmental instruments, 'remarkably prescient in its attempt to define its goal'. He suggests that the ICRW takes an approach of 'establishing trip wires for permitted levels of exploitation' and contends that this approach is 'fundamental to all wildlife management' (Friedheim, 2001b: 20–1). In other words, the ICRW, claims Friedheim, incorporates the precautionary principle into cetacean management – although the 'prescience' he claims might be more a later incorporation, certainly in its explicit formulation. That such an approach is necessary can hardly be in doubt. As Birnie puts it,

> the conservation of cetaceans presents more complex problems of regulation and management than hitherto envisaged. Advances in many modern technologies, accompanied by acceleration in economic development to meet human needs, are accelerating the degradation of many marine ecosystems and the cetacean habitats they sustain.
>
> (2003: 103–4)

This also implies that it is important for the whaling debate to be resolved, for reasons of increased understanding of biological diversity.

Jacobson acknowledges that 'international agreements that serve as charters or constitutions for regulating bodies can and should be interpreted flexibly as their work and its context evolve over time'. He argues, however, that such flexibility must be limited by the objectives and purposes of the organisations created by the agreements, or there is a danger that 'they become anything the majority or politically powerful members want them to become, a clearly unacceptable philosophy'. He concedes that there always will be dissents; but argues that 'reinterpretations that ignore or counter the instrument's objectives and purposes are not acceptable adjustments but are instead violations'. The IWC's actions, he continues, should be seen as 'ultra vires' and 'of no effect at all' in such cases – giving as examples the Southern Ocean sanctuary and a permanent moratorium on commercial whaling (Jacobson, 2001: 91–4, 94–5).

This clearly borders on the nonsensical, to be blunt. Where Jacobson argues that the IWC's actions are ultra vires and therefore not binding, he ignores the fact that the IWC has acted properly in its voting and administrative procedures. When he says that certain actions of the IWC 'should be ignored' (stating of the Southern Ocean sanctuary that it 'can and perhaps should be ignored' (Jacobson, 2001: 94–5)) this is merely wishful thinking. Further, Jacobson's view of the causes of the gridlock within the IWC is only one view of such causes – there are opposing views.

Without a dispute settlement mechanism being provided for in the treaty, is there a way to decide which legal interpretation is the correct one? Could the ICJ make a decision? In 2010, Australia lodged a dispute with Japan in this forum, seeking to have Japan's scientific permit whaling declared in contravention of the ICRW, CITES and the CBD. The litigation is currently ongoing.

Jacobson asks the question whether there is such a thing as the rule of law in the international community, then points out that most states conduct their international relations as though the rule of law does exist. He states that the US has 'been a leader in the IWC coup that is ignoring the clear mandates of the ICRW' and argues that 'without US support and economic power, the anti-whaling states that have packed the IWC would probably not have been nearly as successful as they have been in getting away with their violation of the limitations of the IWC's authority' (Jacobson, 2001: 97–8). Perhaps what he does not give enough credit for here is the *realpolitik* of international law. Every state *does* try to get whatever it wants, and is loath to be bound by international law (or indeed any infringement on its sovereignty) but is unwilling also to be seen to be in breach of international law – and this goes for the US also. While it might attempt to use its muscle to get the result it wants, the US will never do so without trying at the same time to justify itself in the terms of international law. And this will, to some extent at least, restrain the most powerful state – at times, it will be compelled to negotiate in good faith, and to yield in certain areas to get what it wants in others. Jacobson also ignores the fact that, while the anti-whaling states may have been 'getting away with violations', the pro-whaling states have likewise been getting much of what they want.

As flexible as the ICRW itself might arguably be, it soon became apparent that its parties were not easily going to show similar flexibility. I put the issue of the ICRW's purpose to a number of people who are, or have been, involved in the debate. Asked for his view of the purpose of the IWC, past and present, Herman Oosthuizen suggested that this is 'basically the whole debate', and that there are two ideologies – one view led by Japan – seeing the ICRW as being clearly for consumption, with 'biologists' thinking that it should be more for 'conservation dealing with whales'. Both groups, he said, are currently trying to make use of the treaty as it stands, but that it is 'full of loopholes' and 'too old now at the moment' (Oosthuizen, pers. comm., 2006).

When I put it to Horst Kleinschmidt that the ICRW clearly began with a utilitarian goal, and that to argue otherwise would be to give too much credit for foresight to the parties in 1946 when conservation was not a major issue, he answered:

> correct, that's absolutely right, and which side of the debate you stand on informs what you think the purpose of the ICRW is. Obviously many argue that it's as it was in 1946 and they remind the IWC every year of that as the purpose.

His own view, he then explained, was that the IWC is an 'inappropriate instrument in the twenty-first century' for dealing with any form of management of ocean species and that it 'really needs an overhaul'. Asked whether he thought that there are problems with the treaty itself, he said that he believes the treaty to be out of date and to 'reflect a moment in time where a small group of nations got together because either they were directly or indirectly involved in whaling' (Kleinschmidt, pers. comm., 2006).

Best suggested to me that the treaty has a 'grey area' where it is not clear whether conservation is its aim, and that it is a 'weakness' that conservation has not been better defined (Best, pers. comm., 2006). Butterworth said that different countries have different perceptions and that the Japanese are extremely literal in their interpretation; whilst other countries say that the law should evolve. He felt, however, that if conservation includes sustainable use, then it is inherently present, as you cannot have 'orderly development of the whaling industry unless' there is conservation of the animals at the same time, meaning that there is 'really no fundamental conflict there – it's political' (Butterworth, pers. comm., 2006).

Greg Donovan said that he felt that the nature of the ICRW had not changed, and that it is not the Convention itself that is problematic. Whether there are any problems indeed, in his view, depends on the observer's standpoint. 'A tricky question', he called it (Donovan, pers. comm., 2007). Asked a similar question and, in particular, whether the Berlin Initiative of 2003 (which saw the creation of the IWC's Conservation Committee) had changed the ICRW's purpose, Nicola Grandy said that the purpose as set down in the Convention itself had clearly not changed. But, she added, 'what it means to different countries has probably changed over time'. In regard to the Conservation Committee, she said 'we were doing lots of conservation work anyway, we just didn't have a committee, and the Conservation Committee hasn't really changed that' (Grandy, pers. comm., 2007).

The 1982 moratorium came into effect in early 1986. The politics surrounding the moratorium when it came into force are convoluted and will probably never be understood with certainty. Per Watson, the US had threatened to implement the Packwood-Magnuson Amendment, which, if implemented, would probably force compliance by all whaling nations 'under threat of crippling sanctions on their fish exports to the US'. 'What we didn't know', writes Watson, however, 'was that the Reagan Administration had cut a secret deal with the Japanese in December 1984 according to which the Japanese were given a green light to proceed with whaling until 1988' (Watson, 1994: 108).

Komatsu and Misaki write that it is 'generally believed that the IWC banned commercial whaling because whales were nearly extinct' and call this a myth that has become commonly accepted around the world (2003: 19–20). This charge is disingenuous. It may be that, today, such a claim would be a myth; in the early 1980s, however, there *was* reason for genuine concern. One obvious question arises: Why, if the Japanese scientific authorities did not believe that whales were endangered, did Japan agree to be bound by the moratorium? Norway, after all, did not. No answer can be definitive; however, most analysts do seem to believe that Japan bowed to pressure from the US.

In Western countries at least, there is perhaps little awareness that whaling has never actually been *banned*, but has been halted under a moratorium ostensibly temporary in nature. Although due to be reconsidered in 1990, the moratorium has not been so reconsidered and – well past the first decade of the twenty-first century – remains in place, despite various efforts which have been made to break the impasse within the IWC.

Arguably, there was very little progress made in resolving the major issues until IWC 49 in 1997, when the commissioner for Ireland, newly elected as the IWC Chair, attempted to broker compromise. In essence, his proposal was to allow limited and local whaling to take place inside the 200-nautical-mile EEZs of coastal states. Friedheim comments that Ireland 'tried to bypass the core problem of completing the Revised Management Scheme' (2001b: 36). The effort was not, however, successful. Andresen states that the premise for the 'Irish proposal' was that the anti-whaling states would, in principle, have to give the most as they have to accept that commercial whaling can be conducted, implying a fundamental change of policy; but whaling states would have to abandon hopes of a 'real revitalization' of the whaling industry (Andresen, 2001: 248). Strangely, what the anti-whaling states would have to give would be more of a moral victory than anything else for the pro-whaling states; economically, the latter would certainly, if the proposal had been adopted, have found themselves taking far fewer whales than under either the present moratorium or the Revised Management Procedure which has been approved and which would, in force, replace the moratorium.

Linkage between the IWC and CITES is no invention of mine. Both inherently, as multilateral environmental treaties, and explicitly, as directly correlative, the two are linked. Lyster points out that Article XIV of CITES states that the Convention is not intended to affect the provisions of other international treaties, including those pertaining to 'customs, public health, veterinary or plant quarantine'. In a clear reference, he says, to the ICRW, Article XIV.4 relieves parties, where these parties are also party to any other treaty affording protection to 'marine species', of obligations 'imposed by CITES with respect to trade in species in Appendix II which are taken in accordance with the provisions of such other treaty' (Lyster, 1985: 275–6). Despite this, shortly after the creation of CITES, it seems that the IWC became concerned about possible overlap of their areas. The IWC therefore sent a resolution to CITES, seeking to assert the primacy of the IWC in regard to cetaceans and requesting support from CITES. In 1979, CITES, at CoP (Conference of the Parties) 2, responded by acknowledging itself to be a supporter of the IWC. This support appeared in three different resolutions, none of which have been repealed (Gillespie, 2002). This communication is an important point of convergence between the two treaties.

Komatsu and Misaki reiterate that the moratorium was to be reviewed by 1990, and argue that 'both Japan and Iceland believed this promise' and that 'Japan also believed that commercial whaling would be restored after the review'. They conclude that 'so long as the anti-whaling bloc dominates the IWC with the majority votes, it is unlikely that this promise will be fulfilled' (Komatsu and Misaki, 2003: 91). The lack of trust between IWC members is a major stumbling block to resolution of their impasse. Neither pro-whaling nor anti-whaling members trust each other – and, objectively, it would appear that *both* sides have some justification for this mistrust.

Friedheim advises, on the moratorium, that it was imposed to 'provide a pause until a new plan was developed to put whaling on a more scientifically acceptable basis', but it has become a permanent ban *de facto* as the plan (of management measures for implementation) needed to lift it – the Revised Management Scheme – has not been approved. He then comments on the possible implications of this in international law, writing that 'if the moratorium is sustained *de facto* or *de jure*, it must be considered a watershed event, a rare example of a fundamental shift in nature and purpose of an international resource management regime' (Friedheim, 2001b: 7).

Stone comments that it is possible that the 'truculence within the IWC is beyond the balm of any institutional therapy'. He argues that 'the more the issues are conceived as ethical and symbolic, rather than empirical or pragmatic, the less pliant the disputants' and that 'to a large extent, national positions have been hardened – or, more accurately, captured – by inflexible domestic constituencies'. In his view, the basis for what he describes as an 'abdication of policy by central governments is that for the general public in most countries – including the US and Japan alike – whaling is not a high-salience issue' (Stone, 2001: 277). The consequence of this is that those few people for whom it is an important issue are able to exercise a disproportionate influence. A further consequence is that states are able to gain domestic kudos without doing very much for it. As Stone puts it, 'for US administrations, charged with foot-dragging on climate change and biodiversity, concession to whaling opponents is a cost-free sop' (ibid.). There is not much that is hopeful to be gleaned from this assessment. The charge is that members are acting, and will continue to act, only in their own interests – unless the matter is of little interest to them.

This is not a new summation of the position. Writing in 1973, Burton suggested that it is 'exasperating' to consider that 'over the last three decades' the IWC's Scientific Committee has accumulated and analysed much information on the biology of whales, but whaling nations have refused to concur with the accumulated evidence, to reduce quotas and to allow whale populations to recover (Burton,

1980: 171–2). So, the charge has been made from both camps, at different times, that scientific evidence is ignored, and that states continue to act only in their own interests.

3.5 The IWC and CITES

At IWC 46, 1994, a proposed resolution on international trade in whale meat and products was considered. The proposal was made by Argentina, Australia, Brazil, India, Monaco, New Zealand and the US. Japan, however, reminded the meeting of its reservations in CITES, and Norway stated its view that trade was not within the mandate of the IWC. Sweden and Denmark supported the latter view, with Denmark arguing that CITES is the appropriate forum, not the IWC. Voted on, the resolution passed with 14 votes in favour, six against, and seven abstentions (IWC 46, Chair's Report). Venezuela argued in the plenary session that the IWC was 'at a crossroads' but that the crossroads was 'not a choice to hunt whales or not to hunt whales'; rather, the questions were 'can we share the planet's natural sustainable resources in peace and goodwill? Can we respect each other's cultures and backgrounds?' (IWC 46, Verbatim Record).

Of relevance for the present discussion is a submission made by the commissioner for Grenada which is worth quoting at length:

> the resolution before us [on the proposed Southern Ocean Sanctuary] has gone into so much horsetrading without a scientific basis. Because the proponents of the original resolution wished to secure certain countries' support they have taken a purely political decision to move the boundary outside certain EEZs. No scientific basis. My delegation's proposal for the culling within the sanctuary has not been given any thought because it may involve the taking of limited stock of abundant species identified as the minke whales. It is always very dangerous to embark on a sanctuary without making provision for the taking of any species which might cause an imbalance of the ecological mix. Mr Chairman, we know too well that once an amendment is made to the Schedule there is no possibility of turning it back. I must mention that there are dangers in closed areas. Only two nights ago I heard from the observer of Zimbabwe, who is head of the National Parks, and I heard that 13% of Zimbabwe's territory are under National Parks. They have enough space for 50 000 elephants but, because of CITES regulations, they cannot take any and they have 75 000 elephants. What are we doing to poor developing countries? In Zambia you have the same problem with crocodiles, an endangered species. Because of CITES regulations crocodiles have been stopped from being taken. They have rebounded. They have consumed all the fish in the lakes and now they are consuming fishermen. These are true facts. We know what's happening to seals around the world. I've said so before in my intervention. Now we have whales. We may very well find ourselves ending up with a sea of whales and no fish. That is going to have an impact, not on the developed industrialised

> world, but on developing countries, and that's why I have great concern when we establish a sanctuary without scientific basis.
>
> (IWC 46, Verbatim Record)

This is an early example of two important arguments being raised – first, the linkage between whales and elephants in international environmental discourse; and, second, the misunderstanding that it is *because* of CITES that elephants have become 'too numerous'.[15] In 1994, it was only five years since CITES had listed the elephant on Appendix I and the listing could hardly be blamed.

At IWC 47, in 1995, Japan and Norway put forward a proposed resolution on Scientific Advice to CITES. This proposed that the IWC:

> DIRECTS the Scientific Committee, to assist contracting governments by providing technical assistance in: (1) the preparation of proposals to amend the appendices concerning relevant cetacean species; (2) the review of the effectiveness of the inclusion of relevant cetacean species in the appendices.
>
> (IWC 47, Draft Resolutions)

The Netherlands, however, argued that the draft failed to recognise the 'well-established relationship between the two organisations whereby CITES had agreed to reflect IWC decisions in its Appendices and automatically sought IWC advice on any proposals involving the relevant species', and that the resolution would confuse the relationship between CITES, the IWC and individual IWC members. Australia, Germany, Ireland, New Zealand, the UK and the US supported these comments. Denmark 'welcomed the spirit of enhancing collaboration', but argued that the resolution was not in conformity with the CITES criterion of 'taking into consideration the view of intergovernmental organisations with competence for the management of the species concerned'. Japan contended that there was 'no contradiction' with Denmark's view and suggested that, as the moratorium is not based on science, it was 'not appropriate to impose these ideas on another organisation'. St Lucia 'wondered if the IWC has to get involved', while Russia sought 'clarification on whether this was a request from CITES for technical assistance', which it was not. Put to a vote, the resolution was not adopted (five for, 18 against, with seven abstentions) (IWC 47, Verbatim Record).[16]

Also in 1995, the US proposed a resolution, co-sponsored by Brazil and Oman, on improving mechanisms to prevent illegal trade in whale meat. The resolution requested that countries adopt measures which will enable them, by means of

15 It is not because of the CITES Appendix I listings that species might become overpopulated. There is nothing contained within an Appendix I listing that prevents a party to the CITES from culling, or otherwise controlling, a population.

16 In favour, Japan, Korea, Norway, Russia and the Solomon Islands; against, Antigua and Barbuda, Australia, Austria, Brazil, Chile, Finland, France, Germany, India, Ireland, the Netherlands, New Zealand, Oman, South Africa, Spain, Sweden, the UK and the US; abstentions, China, Denmark, Dominica, Mexico, St Lucia, St Vincent and Switzerland (IWC 47, Verbatim Record).

disposing of stockpiles, testing markets and identifying whale meat by DNA or isozyme analysis, to prevent whale meat from being 'taken in contravention of the ICRW and CITES' – and prohibiting the sale of meat 'not taken in accordance with ICRW and CITES provisions'. Switzerland was not opposed to the resolution, but stated that it believed the resolution went 'far beyond the scope of the ICRW, while CITES has the competence in international trade' – it therefore indicated that it would abstain. According to Switzerland:

> Article I of the [ICRW] clearly states that the Convention applies to factory ships, land stations and whale catchers under the jurisdiction of the contracting governments and to all waters in which whaling is prosecuted by such factory ships, land stations and whale catchers. Full stop. It does not apply to stockpiles of whale meat or other parts such as baleen or bone which are sold to tourists, for instance in the United States and Canada, nor to internal markets, nor international trade. International trade is in the competence of CITES where this matter has and will be taken care of appropriately.
>
> (IWC 47, Verbatim Record)

Japan, voicing similar concerns to Switzerland, was likewise opposed to the resolution, and pointed out that there had been no preceding discussion at this meeting about monitoring domestic markets. In debate, Japan's commissioner argued that 'this drafted resolution intended to do something, requests something, for all government, not only the IWC member nations but also all other governments or other entities, and this is clearly going beyond the competence of IWC also' (IWC 47, Chair's Report).

Norway, Russia and St Vincent and the Grenadines voiced similar concerns to those of Switzerland. Denmark suggested a minor change of wording, which was agreed to by the US, while Chile, the Netherlands and Antigua and Barbuda suggested that they would support the resolution. St Lucia expressed concern for starving people, to the US' suggestion that stockpiles could be destroyed; in response the US explained that 'disposal domestically of stockpiles could be carried out by any means'. The resolution was then adopted: 21 for, three against, with six abstentions. Mexico explained that it had abstained because of 'the issue of cooperation between international organisations on trade' and argued that this provided 'another reason for updating the ICRW'. Japan, finally, stated that it 'will continue its efforts to prevent smuggling of whale products', but that 'because of the legal issue and potential conflicts within the Commission' it was opposed to the resolution (IWC 47, Chair's Report).[17]

17 In favour were Antigua and Barbuda, Australia, Austria, Brazil, Chile, Denmark, Finland, France, Germany, India, Ireland, Rep of Korea, Monaco, the Netherlands, New Zealand, Oman, South Africa, Spain, Sweden, the UK and the US; against were Japan, Norway and the Solomon Islands; abstentions were China, Dominica, Mexico, St Lucia, St Vincent and Switzerland.

After the vote had been taken, Japan's commissioner noted that Japan 'has been made (*sic*) and will continue its efforts to prevent smuggling attempts in whale products'. He then described the situation as 'rather unfortunate' and suggested that this was because:

> one country is willing to make a spontaneous *bona fide* cooperation even knowing the difficulty of the scope of the Convention, but the proposed resolution would eventually expose an unnecessary conflict among the Commission and I think our government will reconsider the cooperative position to this Commission.
>
> (IWC 47, Verbatim Record)

At IWC 48, 1996, Austria, Brazil, New Zealand and the US proposed a resolution 'on improving mechanisms to restrict trade and prevent illegal trade in whale meat', calling on member countries, and other entities, to report on stockpiles of meat, relevant domestic laws governing sale and possession, enforcement actions, and calling for support for IWC recommendations and CITES obligations. Japan opposed the resolution on principle as being, according to Japan, outside the Commission's competence. Norway took a similar stance. Put to a vote, the resolution passed (20 for, seven against, two abstentions). Denmark explained that it had voted 'no' because, although it supported the general intention, in its opinion the matter was for CITES. Switzerland abstained, as it thought CITES to be the competent body and that domestic markets are outside of the scope of the ICRW (IWC 48, Chair's Report).[18] In plenary debate, Japan's commissioner had argued that:

> any market control is outside of the competence of the Convention and in this sense this resolution could challenge the right and duty of sovereign states and also we are not in the position to acknowledge the listing status of the large whales at the CITES Appendices.
>
> (IWC 48, Verbatim Record)

It was already clear, therefore, that any overlap between the two treaties was going to become contentious.

At IWC 48, Zimbabwe, as an observer, submitted a (written) opening statement. Zimbabwe said that although it is a landlocked state, it was 'grateful to the IWC for the lessons learned during this and previous meetings from the discussions held on conservation issues'. Zimbabwe stated that it 'recognises that wild life is a unique economic resource upon which a growing global industry is developing'. The

18 The vote ran: for, Australia, Austria, Brazil, Chile, China, Finland, France, Germany, Ireland, Korea, Mexico, Monaco, the Netherlands, New Zealand, Oman, South Africa, Spain, Sweden, the UK and the US; no, Denmark, Grenada, Japan, Norway, St Lucia, St Vincent and the Solomon Islands; abstentions were Russia and Switzerland.

advantage of this particular industry, according to Zimbabwe, lies in its 'ecological sustainability in comparison to other land uses and, for Zimbabwe, its competitive advantage in world markets'. Zimbabwe explained that, 'in accordance with its commitment to conservation and its resolve to promote enhanced sustainable rural prosperity and a more equitable apportionment of the benefits from proper use of the nation's wild life resources', it continues, inter alia, to 'participate in those international treaties and conventions which are consistent with Zimbabwe's policies for conservation and sustainable use of wildlife'. Zimbabwe then explained that some animals which were once thought to have been going extinct have recently shown 'spectacular recoveries'. 'A case in point here', ran the statement:

> is the elephant which increased to more than 80 000 by 1980. Today due to the ban on trade in ivory products, the elephant population continues to expand but to its own detriment as it destroys large areas of vegetation and comes into more conflict with the rural people because they continue to destroy crops and even kill people at times. Under these circumstances, people take the law into their own hands and start killing these elephants illegally as they are not benefiting from these animals. This only erodes the policy of sustainable utilization that Zimbabwe is attempting to promote, under which rural peoples have the authority to manage their wild life and other natural resources and benefit directly from doing so.

In 1997, concluded the statement,

> Zimbabwe will host the 10th CITES meeting in Harare and will promote the principle of sustainable utilisation of renewable resources as an important conservation tool and hopes to encourage all international conservation bodies to adopt the same principle along the same lines.
>
> (IWC 48, Opening Statements)

At IWC 49, in 1997, the IWC considered the results of the CITES CoP 10, held in Harare in June of that year. Japan, in its (written) opening statement, said that it would, during the meeting, 'make every effort to accommodate whatever reasonable requests are made of us. We are very willing to make whatever compromises are necessary to ensure the safety of whale populations and relieve the anxieties of those opposed to whaling'. Japan cautioned, however, that it could not 'betray the principles upon which this organization was founded, nor the principles of other international compacts such as UNCLOS, Agenda 21, CITES, and the Kyoto Declaration'. 'Please remember', asked Japan,

> that at the last CITES meeting in Zimbabwe, a majority of nations voted to allow the sustainable use of abundant whales such as minkes. And 53 nations – almost twice the number of nations opposed to whaling at the IWC – voted to downlist the abundant Antarctic minke whale populations and allow

> sustainable use. The CITES vote made it very clear that most of the people in the world would approve of whaling if it does not endanger the whale populations.

The IWC, suggested Japan, has the responsibility to ensure that this does not happen (IWC 49, Opening Statements).

Norway, in its (written) opening statement, said that there had been 'clear international recognition of science as the basis for conservation discussion at the CITES meeting in Harare in June this year', where a 'majority voted in favour of down-listing the Northeast and Central Atlantic minke whale stocks, although less than the two-thirds majority needed to change the listing'. Norway then stated that it was 'convinced that there will be further progress' in respect of whales at the next CITES meeting, 'as the principle of sustainable use is increasingly gaining support internationally'. Norway then explained that it wished to underline 'the division of competence between CITES as a trade organisation and IWC as a management organisation' (IWC 49, Opening Statements).

In plenary debate, Japan indicated that it wished to discuss the CoP 10 'intensively'. Japan's commissioner then described the CoP as a 'historical moment in the period of international conservation of wildlife resources'. Japan suggested that the CoP had seen two major developments: first, 'that nations had moved from a position of total protection to one of sustainable use to achieve their conservation objectives'; and second, that the sovereign rights of nations had been recognised. Japan suggested that the downlisting proposals for gray and minke whales had shown 'a dramatic change in the level of support made possible by the expression of sovereign rights through the use of the secret ballot'. 'National policies', Japan argued, 'were not dictated by interest groups and differences of culture and ethics were respected'. Japan argued that the 'overall result of CoP 10 dealing with the enforcement of sustainable use' was evident from the number of proposals to downlist 'so-called charismatic' species, such as 'whales, elephants, sea turtles and rhinos'. Japan described the downlisting of the Botswana, Namibia and Zimbabwe elephant populations as an 'indication of this change of attitude within the state parties' (IWC 49, Verbatim Record).

The Japanese commissioner pointed out that, in respect of whales, four proposals for downlisting were voted on: three by Japan and one by Norway. Japan's last proposal 'received 53 votes in favour and 59 against'. 'The following proposal coming from Norway', he continued, 'received a majority vote of 57 in favour, 51 against'. 'These votes', he suggested, 'constitute a drastic change compared with CoP 9 in 1994 when the Norwegian proposal received a low level of support'. The commissioner proceeded to argue that it had been the use of the secret ballot at CITES CoP 10 that had 'helped the governments in voting in that election in line with their national interest in the philosophy' since they could, in this way, avoid 'appropriate (*sic*) pressure from other governments and extremist groups'. The secret ballot, he argued, 'has in fact been the total needed to achieve greater sovereignty'. The Japanese commissioner then argued that:

> some very important messages sent to the international community by CoP 10 were, among others, the following. Sustainable use shall not apply only to chickens and cows but as well to all species of plants and animals of the world. Developing countries will not accept to have their conservation policies dictated any longer or enforced by so called interest groups. Interference of so-called interest groups into the affairs of other sovereignty issues shall not take place from now on. Difference of the culture, tradition, ethics and values shall be respected by all. In reality, Mr Chairman, the CITES community which is more than three times larger than the IWC memberships including more than ninety five per cent of the membership of the IWC is to represent the world opinion and CITES itself has come from dealing with species to caring for people and species. It has moved from applying blanket prohibition to seeking solutions for conservation problems. CITES has finally espoused the agreement of the UN Convention, UN Conference on Environment and Development and recognised sustainable use of living resources as a means to solve conservation problems.
>
> (IWC 49, Verbatim Record)

It is illuminating to see that both Japan and Norway sought to bring back lessons from CITES CoP 10 and apply them to the IWC. According to the Chair's Report, Japan 'emphasised the two points made by the IWC observer, which were agreed by the Commission', that, first,

> CITES parties still see value in keeping the links with IWC but the IWC needs to complete its work on the RMS sooner rather than later, otherwise the scientific basis by which CITES has opted for operating will become frustrated

and that second, 'the IWC Secretariat should continue to develop strong links with the CITES Secretariat, and the Scientific Committee should maintain links with the Animals Committee of CITES on relevant issues' (IWC 49, Chair's Report).

The commissioner for the US, however, responded that 'some delegations seem to have been comforted by minor changes in the voting record at the recent CITES meeting'; whereas all downlisting proposals for whales had been rejected and that, in his opinion, was what was really 'reflective of the true international opinion'. The US also pointed out that, at CoP 10, a proposal to rescind CITES Resolution 2.9 (which had acceded to the IWC's request to CITES to recommend to its parties that they refrain from issuing export or import permits for any species protected by the IWC from commercial whaling) was defeated 'overwhelmingly' by a vote of 51 (against) to 27 (for). Norway responded, however, by arguing that the US' point should not be interpreted to give the impression that there is a 'cross link' between the conventions, such that CITES cannot downlist species unless the IWC moratorium has been changed. Norway pointed out that the resolution referred to by the US 'is a recommendation to governments to follow the practice of not trading while the moratorium' is in place. Norway then highlighted the fact that

'CITES over the last couple of years have established new criteria' for listing of species, and that these criteria are based on the sustainable use concept, with science as the basis, which makes it 'quite clear that CITES on its own basis according to its criteria can list species on list one or two' (IWC 49, Verbatim Record).

Of other countries, Brazil argued that the IWC is the 'recognised and appropriate forum to decide upon matters relating to the conservation of cetaceans', and that, therefore, the Commission 'has authority over issues such as the trade in whale meat, in so far as it directly affects whale conservation'. This view has been expressed before; a point in support which it overlooks is to mention that the United Nations Convention on the Law of the Sea (UNCLOS) actually mentions 'appropriate managing authorities' in respect of cetaceans, and that the IWC has long been considered the most appropriate such authority. Australia's commissioner said that he thought that 'all those who attend the meetings of that sort develop their own interpretation and I wouldn't say that the Australian interpretation equates with that in Japan'. He suggested that, from Australia's perspective, the 'most significant outcome of the Harare meeting was the overwhelming vote in favour of maintaining close relations between this Convention and the CITES'. St Lucia's commissioner then suggested that it might be 'time to renegotiate this Convention where Governments interested in the global moratorium on whaling and a global sanctuary for whaling would go their way and a new IWC would emerge, one which is interested in regulating whaling'. What the world needs, continued the commissioner for St Lucia, is 'to take the global view, the new global view, of sustainable utilisation in stride' (IWC 49, Verbatim Record). While opinions differed on the degree of relevance for the IWC of the decisions made at CITES CoP 10, what is not in doubt is that the pro-whaling IWC members sought to use it as a precedent in support of their arguments in the IWC.

Also in 1997, the Commission passed a resolution on Improved Monitoring of Whale Product Stockpiles.[19] Japan's commissioner, in plenary debate, opposed this resolution on the basis that the IWC does not have 'a management capacity' over the issue. Denmark's commissioner then added that Denmark has 'a little concern about this whole exercise about the competence between IWC or World Trade Organisation and CITES'; later adding more definitely that his country felt that the area was 'the competence of CITES, the competence of the World Trade Organisation' (IWC 49, Verbatim Record).

At IWC 50, in 1998, in the Technical Committee, Norway, with Japan in support, argued that discussions of stockpiles of whale products and trade questions are 'not within the scope of the Convention'. Japan 'noted the adoption of a decision on trade in whale meat adopted at' CITES CoP 10, and both Japan and Norway stated that they were 'willing to discuss such matters in what they

19 The vote went: for, Australia, Austria, Brazil, Chile, China, France, Germany, Ireland, Monaco, the Netherlands, New Zealand, Sweden, Switzerland, the UK and the US; against, Antigua and Barbuda, Dominica, Grenada, Japan, Norway, St Lucia, St Vincent and the Solomon Islands; with abstentions being Argentina, Denmark, Finland, Korea, Mexico, Oman, Russia, South Africa and Spain.

considered to be the appropriate fora (WTO and CITES)'. In the Commission itself, the Netherlands, supported by the UK, voiced its concern over illegal trade and urged thorough investigation and action; but Japan 'reiterated its view that matters of trade and its domestic market fall outside the IWC and are under its [Japan's] own sovereignty, although it cooperates with CITES in international, and TRAFFIC Japan on domestic, issues' (IWC 50, Chair's Report).

The US, with Australia, Brazil, Monaco, New Zealand and Oman, proposed a resolution on continued cooperation between the IWC and CITES. The focus of this resolution was particularly on the problems of illegal trade in whale products. Denmark and the UK supported the resolution, but felt that they could not co-sponsor it, as the European Union 'is responsible for their trade affairs'. Norway did not give support as it felt that the resolution trespassed on the competence of other organisations; Japan expressed its view, as before, that the issue is outside IWC competence – and also that some of the resolution's paragraphs were 'contrary to the facts'. Japan, seconded by Antigua and Barbuda, proposed a number of amendments. After several drafts and revisions, the resolution went to a vote and was adopted by 23 votes for, seven against, with three abstentions (IWC 50, Chair's Report).[20]

20 The Resolution on Cooperation Between the IWC and CITES reads as follows:

> WHEREAS it is the purpose of the Commission to provide for the effective conservation and management of whale stocks; WHEREAS the Commission is the universally recognised competent international organisation for the management of whales and whaling; RECALLING that the IWC has passed resolutions welcoming the continuing cooperation between CITES and the IWC on the issues related to trade in whale meat, and urging all governments to continue to support IWC and CITES obligations with respect to this issue; WHEREAS the [COP of CITES] recognised in Res. Conf. 2.9 the mandate of the IWC, and requested that the CITES Secretariat consult with the IWC concerning proposals to amend the CITES appendices; ACKNOWLEDGING that all species of whales in the Schedule to the IWC have been listed in Appendix I by CITES (with the exception of the West Greenland stock of minke whales which is listed in Appendix II by CITES) pursuant to and in recognition of the establishment of zero quotas for commercial whaling agreed by the Contracting Governments to the IWC, and other decisions of the IWC relating to the status of great whale species; WHEREAS CITES has recommended that Parties not issue any import or export permits for trade in any whale stocks for which the IWC has set zero catch limits; WELCOMING the recent decision by [the 10th CoP of CITES] to uphold Res. Conf. 2.9; NOW THEREFORE the Commission: EXPRESSES its appreciation to the [CoP of parties] to CITES for its reaffirmation of the relationship between CITES and the IWC; RECOGNISES the important role of CITES in supporting the conservation of whale stocks and the IWC's management decisions, and reaffirming the importance of continued cooperation between CITES and IWC; REAFFIRMS the need for Contracting Governments to observe fully IWC resolutions (IWC Resolutions 1994-7; 1995-6; 1996-3; 1997-2) addressing trade questions, in particular with regard to the problem of illegal trade in whale products; URGES all governments to provide the information specified in previous resolutions to the Commission; DIRECTS the IWC Secretariat to forward this resolution, and this year's reports of the Infractions Sub-committee to the CITES Secretariat.

The US had introduced the resolution by welcoming the recent decision of CITES CoP 10 to uphold Resolution Conference 2.9 in which parties to CITES recognise the mandate of the IWC. That resolution, according to the US, 'solidifies a relationship that began in 1978 (*sic* – probably 1977 was meant) with an IWC resolution requesting that CITES member states take all possible measures to support the IWC ban on commercial whaling for certain stocks and species'. Since that time, the US continued, 'CITES has taken numerous actions in support of IWC initiatives including recommending that parties not issue any export or import permits for trade in any whale stocks for which the IWC has set catch limits'. The US recommended that it was therefore important to recognise CITES important role in supporting the IWC – 'and to reaffirm the importance of continued cooperation between the IWC and CITES'. Norway, however, argued that 'Resolution 2.9 is simply a reference to individual governments not to issue export permits'. 'It does not', Norway said, 'set up the sort of relationship between CITES and the IWC which is claimed in the resolution and which was referred to in the US statement'. The Japanese commissioner was firmer, arguing that the resolution 'actually trespasses or it is outside the competence of the IWC' (IWC 50, Chair's Report).[21]

At IWC 50, Zimbabwe, as an observer, again submitted a (written) opening statement. Zimbabwe stated that it had, for many years,

> supported the scientific findings dealing with the conservation of the whole ecosystem and not those dealing with the protection of only a specific or charismatic species: the disastrous imbalance then created to the ecosystem has been experienced on numerous occasions in different parts of the world and needs to be avoided.
>
> (IWC 50, Opening Statements)

This is an interesting example of Zimbabwe picking up on the way in which the 'ecosystem approach' has recently been reinterpreted by the pro-whaling group to support its arguments for resuming commercial whaling.

There was increased interest in the relationship between CITES and the IWC at IWC 51 in 1999, where a resolution on cooperation between the IWC and CITES was introduced by the US. The resolution 'commented that all species of whales', apart from the West Greenland minke (listed on Appendix II), are listed on CITES Appendix I. Voted on, the resolution was adopted with 21 for, ten against and three abstentions (Chair's Report, IWC 51).[22]

21 The vote on the resolution reflected: in favour, Argentina, Australia, Austria, Brazil, Chile, Denmark, Finland, France, Germany, India, Ireland, Italy, Mexico, Monaco, the Netherlands, New Zealand, Oman, South Africa, Spain, Sweden, Switzerland, the UK and the US; against, Antigua and Barbuda, Dominica, Japan, Norway, Russia, St Lucia and the Solomon Islands; with Korea, St Vincent and China abstaining.

22 As adopted, the Resolution on Cooperation Between the IWC and CITES reads as follows:

> WHEREAS it is the purpose of the [IWC] to provide for the effective conservation and management of whale stocks; WHEREAS the IWC is the universally recognised competent

In proposing the resolution, the US commissioner had suggested that 'the IWC and CITES are both interested in whales; while the IWC provides for conservation and management of whale stocks, CITES regulates their international trade', and, further, that as the two organisations provide for different aspects of regulation of the same species, it is 'imperative that they cooperate as closely as possible'. Norway's commissioner, however, responded by suggesting that Norway saw 'this resolution as another attempt to prevent CITES to adhering (*sic*) to its own legal basis, the CITES convention according to which only species or stocks that are

international organisation for the management of whale stocks; ACKNOWLEDGING with satisfaction that all species of whales in the Schedule to the IWC (*sic*) have been listed in Appendix I of CITES (with the exception of the West Greenland stock of minke whales, which is listed in Appendix II by CITES) pursuant to and in recognition of the establishment of zero catch limits for commercial whaling agreed [by the IWC], and other decisions of the IWC relating to the status of great whale species; WHEREAS by virtue of the inclusion of these species in CITES Appendix I and Resolution Con. 2.9, CITES requires that Parties not issue any import or export permits for commercial trade in any whale stocks for which the IWC has set zero catch limits; WELCOMING the recent decision by the 10th [CoP of CITES] to uphold CITES Res. Conf. 2.9; WELCOMING as well the recent decisions of the 10th [CoP of CITES] (Decisions 10.40–10.43) that recognised the need for international cooperation in monitoring and controlling the illegal trade in whale meat; RECOGNISING that the IWC has made progress toward completing the Revised Management Scheme, specifically by the endorsement of the Revised Management Procedure, by the revision of the requirements and guidelines for conducting surveys and analysing data within the Revised Management Scheme, and by the clarification of arrangements to ensure that total catches over time are within the limits that would be set under the Revised Management Scheme; NOW THEREFORE the Commission: RECOGNISES that the IWC management regime prior to the establishment of zero catch limits for commercial whaling led to the global demise of the whale stocks; FURTHER RECOGNISES that the IWC has not completed the necessary measures to ensure that commercial whaling catch limits are not exceeded, that whale stocks can be adequately protected, and that all whaling by IWC member countries is brought under effective IWC monitoring and control; RECOGNISES the important role of CITES in supporting the conservation of whale stocks and the IWC's management decisions, and reaffirming the importance of continued cooperation between CITES and the IWC; RECOGNISES as well the important role of CITES in detecting illegal trade in whale meat through inclusion of whale species in CITES Appendix I; EXPRESSES its appreciation to the [CoP of CITES] for its continuing reaffirmation of the relationship between CITES and the IWC; DIRECTS the Secretariat, when the IWC is requested to provide comments on any proposal submitted by a CITES Party to transfer any whale species or stock from Appendix I to II, to advise the CITES [CoP] that the IWC has not yet completed a revised management regime which ensures that future commercial whaling catch limits are not exceeded and whale stocks can be adequately protected; FURTHER DIRECTS the Secretariat to advise the CITES [CoP] that zero catch limits are still in force for species of whales which are managed by the [IWC]; . . .

In the end, the vote ran: in favour, Argentina, Australia, Austria, Brazil, Chile, Finland, France, Germany, Ireland, Italy, Mexico, Monaco, the Netherlands, New Zealand, Oman, South Africa, Spain, Sweden, Switzerland, the UK and the US; against, Antigua and Barbuda, Dominica, Grenada, Japan, Norway, Russia, St Kitts and Nevis, St Lucia, St Vincent and the Solomon Islands; with China, Denmark and Korea abstaining.

threatened with extinction should be listed in Appendix I', and, harshly, that the 'IWC should not try to impose its own habits of diverting from the legal rules governing its activities on other organisations with independent responsibilities'. Along the same lines, the commissioner for Antigua and Barbuda said that:

> we find that this is sort of confrontational and is tantamount to a subtle interference by the IWC in the affairs of CITES which should never be tolerated, and we think it is a way in which this organisation attempts to direct and influence the work of another independent organisation.

Antigua and Barbuda then argued that the IWC 'has no competence in trade', and that the IWC was 'engaged in an expansionist tendency which is reflective of the type of domination that is being attempted in this organisation by a few countries, a minority of countries' and that the resolution was 'dangerous' (IWC 51, Verbatim Record).

Japan's commissioner said that CITES should make its own judgement based on the scientific information provided by the IWC, but that the 'zero catch limit is a politically decided figure, it is irrelevant, it is not relevant to the resource of stock status', before claiming that 'this kind of action here at this Commission is quite a serious problem which demonstrates the serious nature of the problem encountered by this Commission'. The commissioner from St Lucia then argued that:

> there is a light at the end of the tunnel and this resolution gives me an impression that there is desperation, desperation in the sense that we have to remind CITES that there exists a relationship between the two organisations.

'Try as we may', the commissioner went on, '140 countries next year will take the right decision, they did it a few years ago on the elephant and they know what to do'. Finally, the commissioner suggested, sarcastically, that the name of the ICRW might appropriately be changed to the 'International Convention for the Lock Up of Whales', but that 'next year CITES, an independent body with 140 countries around the world who believe in the sustainable use of living marine resources, will take the right decision'. Denmark, likewise, felt that 'trade issues are the concern of CITES and the WTO' (IWC 51, Verbatim Record).

Australia explained that 'at the last CITES meeting, Japan had moved a resolution that relationships between CITES and the IWC should be radically changed so that CITES no longer requested the advice of this Commission'. 'Fortunately', the Australian commissioner continued, 'that resolution was defeated by a very large majority', and that subsequent debate within the IWC had concerned whether appropriate information had been given to CITES or not, and that this resolution was intended to clear up such confusion. Further, the commissioner argued that he found it 'intriguing' that Japan was 'arguing strenuously for increased cooperation between this organisation and others in one part of the debate when it would favour their position', but that 'when they see perhaps the cooperation with another organisation may not favour their position, suddenly such cooperation is outside the competence of this body'. Further, he suggested, 'Japan argues, and has argued loudly and long, within CITES that that body should have no jurisdiction in marine fisheries. It argues in this body that CITES is the appropriate organisation to have jurisdiction in whale fisheries' (IWC 51, Verbatim Record).

In response, the Japanese commissioner said that:

> the resolution that was mentioned was submitted by Japan to CoP 10 of CITES in 1997 and we gave the support of 29 votes and unfortunately the resolution failed, but that resolution was intended to strengthen the relationship between CITES and the Scientific Committee of the IWC. Because the bottleneck of the problem of lack of flow of scientific information from the Scientific Committee of the IWC to the CITES is this Commission itself.

The Japanese commissioner then argued that, by analogy, the FAO functions:

> normally in the light of its constitution, as it is promoting sustainable use of resources which is in line with the main objective stipulated in the constitution of FAO, therefore since the FAO has expertise on marine species, marine resources, the FAO should have practical responsibilities for the management of the marine species.

However, the commissioner went on,

> as far as whales are concerned, this very Commission, the IWC is not functioning in accordance with the spirit and the letter of the ICRW, and that normal management of the whale stocks are not expected from this Commission and therefore we think it reasonable to ask CITES to manage whale stocks rather than leaving it up to the IWC.
>
> (IWC 51, Verbatim Record)

In the resolution, the IWC reaffirmed the importance of its relationship with the CITES and advised CITES once again that zero catch limits are still in force for all species of whales managed by the IWC (IWC 51, Resolution 6). It is thus not spurious for a writer like me to link CITES and the ICRW/IWC – the parties and the organisations have themselves wrestled over the links. However, while it is essential to establish as many links as possible between CITES and the ICRW/IWC, for full understanding, it is the specific links between elephants and whales that I am most interested in.

At IWC 51, Zimbabwe again gave a (written) opening statement, in which it said that it 'strongly believes that its philosophy of sustainable use applies equally to its native elephants and to all other living species, including whales, for the greater benefit of people, their respective culture and traditions' (IWC 51, Opening Statements). It is interesting that Zimbabwe, despite having been given the restricted downlisting of the elephant that it had argued for at CITES CoP 10, still considered it useful to use the IWC as a forum to push their views – perhaps debts were owed.

At IWC 52, 2000, Japan asked that the NGO Greenpeace's Observer credentials be withdrawn due to a collision between a Greenpeace vessel and a Japanese research vessel in the Antarctic. It is worth mentioning here simply because the US 'noted that there were conflicting accounts of the incident, and that CITES had taken no action in response to a similar request' (IWC 52, Chair's Report). It is significant, of course, that Japan had tried using both CITES and the IWC as fora for protest against an NGO which had protested against whaling.

At IWC 52, Japan reported that it, together with Norway, had proposed, at CITES CoP 11, the downlisting of gray whales in the eastern North Pacific and of minke whales in the Antarctic, North Atlantic and North Pacific. The Chair (from Ireland) argued that the views which the CITES Secretary-General had expressed were a 'clear signal that the RMS should be finalised'; although the views were 'not necessarily a view of' the CoP itself. St Lucia suggested that it was 'concerned about the perception of the IWC from the outside', and, 'referring to the letter from the CITES Secretary-General', suggested that 'the only reason that whales have not been downlisted is because of the agreement the IWC has with CITES not to do anything to undermine the IWC'. The UK explained that, in its opinion, the letter from CITES 'expressed the views of the CITES Secretariat, not those of the organisation' itself. St Vincent and the Grenadines, however, stated that it disagreed with the UK's interpretation of the CITES letter (IWC 52, Chair's Report).

In plenary discussion, Japan's commissioner argued that the results of the vote on downlisting at CITES CoP 11 'shows that the proposal and contention by Japan and Norway, namely the management of the whale stocks should be based upon the scientific evidence' is 'now well accepted by the many countries of the world'. Japan argued further that the result, of more countries voting for downlisting than there were signatories to the ICRW, showed that the view is wrong that the whaling ban reflects international public opinion. The Japanese commissioner then suggested that:

> by the next CoP of CITES unless we start sustainable commercial whaling with the completion of RMS the international society will consider IWC as having abandoned its responsibility for the management of whale stocks and that IWC is going to lose its credibility.

Brazil expressed the view that 'the synergy between the two organisations maintained the IWC's competence', and argued that attempts to reopen trade were premature and would be 'detrimental to this relationship'. The Norwegian commissioner argued, however, that:

> synergy between the organisations has been a negative one. It is almost, if I may use a popular expression, a Catch-22 situation, where you see the CITES hiding behind the back of the IWC on the issue of transferring the minke whale from Appendix I to Appendix II.
>
> (IWC 52, Verbatim Record)

The commissioner from Switzerland expressed the view that the Chair of the IWC had advised CITES in 1997 in Harare, and again in 2000 in Nairobi, that the IWC was 'very close before closing the work on the RMS'. The commissioner said that he doubted:

> if in the year 2002 when the next CITES Meeting is going to be held the Chairman should be forced to say the same statement again. I don't think the world will understand why this body takes more than seven years to finalise what is now a document of about five pages.

Switzerland's position is not always along the lines that might be expected, given its geographic and political closeness to anti-whaling states. Monaco's commissioner, on the other hand, said that:

> we wish to express our growing unease with the notes, messages and even faxes of displeasure expressed by certain organisations toward IWC. We feel that such expressions are totally inappropriate. They represent undue interference in the business which is that of our Commission and they tend to indicate that the image, we have an image perhaps of weakness, and we should not certainly bow to such pressure.
>
> (IWC 52, Verbatim Record)

Zimbabwe again delivered a written opening statement in which it stated that the meeting:

> comes hardly two months after CITES CoP 11 held in Kenya, where for us, the concept of sustainable utilization of renewal resources gained further acceptance through the retention of our elephant population on Appendix II thus allowing us to locally trade in some of the elephant products, notably meat and skins.

It is unclear what Zimbabwe meant by 'local trade'. Zimbabwe went on to suggest that, 'in an attempt to improve the management of our elephant populations', it was 'currently engaged in the development of Transfrontier Conservation Areas with its neighbours' as a way to promote 'inter-state cooperation in the management and sustainable use of ecosystems that transcend national boundaries with the aim of increasing economic benefits to their populations' (IWC 52, Opening Statements). Zimbabwe did not mention whales or whaling in its opening statement, obviously assuming that the linkage and relevance would be obvious to all. Linkage of species was also a theme in Japan's (written) opening statement, in which it accused Australia of 'double standards' and suggested that if minke whales were to be renamed 'the kangaroos of the sea' the Australian public might support their use. Japan also asked how a 'country that argues it should be allowed to emit more greenhouse gases' can 'object to the sustainable use of whale resources' (IWC 52, Opening Statements).

At IWC 53, in 2001, Iceland sought to rejoin the IWC. However, what caused controversy was that Iceland sought to rejoin with its instrument of adherence conditional on a reservation to the commercial whaling moratorium. Iceland explained that it had withdrawn in 1992 because it had believed that the IWC 'was no longer operating in accordance with the Convention and had become a non-whaling commission rather than a whaling commission'. Iceland, however, now felt that there were signs within the IWC of support increasing for sustainable whaling, and that Iceland had, therefore, rejoined in order to 'have an influence on the discussions taking place'. Iceland claimed that it had the right under international law to adhere with a reservation. After complicated procedural arguments for and against Iceland's contention, the country was not permitted to adhere with a reservation (IWC 53, Chair's Report).

Germany proposed a resolution regarding ongoing whaling occurring despite the moratorium. Inter alia, Germany noted that Norway had decided to resume international trade in minke whales, despite minke whales being listed on Appendix I of CITES. Norway, in response, described it as 'unacceptable' that the resolution sponsors did not respect Norway's right under international law to lodge reservations, even where making such reservations is specified as a right as in both CITES and the ICRW. Norway asked that the proposed resolution be withdrawn – this call being supported by Antigua and Barbuda, Dominica, Grenada, Iceland, Japan and Russia. The US then 'recalled' that Norway had continued commercial whaling despite similar resolutions in the past, and that Norway's present indication that it would re-open trade in whale products was 'contrary to the moratorium and to CITES'. The UK pointed out that Norway's proposal at CITES CoP 11 to downlist certain whale species had not been adopted. Voted upon, the resolution received 21 votes for, 15 against, with one abstention (IWC 53, Chair's Report)

In its (written) opening statement, Norway had advised that it found it:

> highly surprising and unacceptable that Germany, and its co-sponsors, do not seem to respect international law and a country's right to lodge relevant reservations *even* where such action is clearly expressed as a specific right, as it is in CITES and the IWC. It is specifically disturbing that Germany proposes a resolution against another country exercising such rights, as long as it has lodged seven reservations itself under CITES.

'Norway', the statement went on to record, 'has not used its right to issue export permits for minke whale products since 1988', and to argue that a:

> majority of the parties to CITES do respect the fact that the two stocks of minke whales we have proposed for downlisting to Appendix II are *not* threatened with extinction, which is a legal and scientific prerequisite for listing on Appendix I, and have supported Norway at the last two CoPs of CITES.
>
> (IWC 53, Opening Statements)

With regard to CITES and 'the need to improve mechanisms for the transfer of scientific samples between countries', the Scientific Committee urged member nations at IWC 53 to 'consider nominating certain "centres" to be given institutional CITES permits to facilitate the import and export process' (IWC 53, Chair's Report).

3.6 Entering the twenty-first century

IWC 54, 2002, was hosted by Japan, and the port city from which the whaling fleets leave – Shimonoseki – was chosen as the venue. The meeting witnessed some of the most determined efforts yet made to resume commercial whaling – according to the Japan Whaling Association, the meeting closed 'with its stalemate remaining but pro-whaling groups gaining steady ground in voting' (*JWA News*, 2002).

Iceland again deposited an instrument of adherence – with a reservation. Iceland stated that it now regarded itself as a member of the IWC, despite the events and the Chair's ruling of the previous year. After argument as complicated as the previous year, Iceland was again not admitted (IWC 54, Chair's Report).

While discussing the aboriginal quota in Greenland, the UK raised concerns about 'the practice of whale products being exported to Denmark from Greenland', believing that the practice 'contradicted the philosophy that aboriginal subsistence products must be consumed locally'. Denmark, however, explained that the practice is 'in accordance with CITES' and 'is considered a transfer within the Kingdom of Denmark, not an export'. The UK responded by expressing concern at the precedent that might be set. Later in the meeting, the UK did argue that, only a few days before, the Environmental Investigation Agency NGO 'had purchased a package of whale meat in Nagasaki labelled as coming from Greenland'. The Danish commissioner and a representative of Greenland's government, however, referred to CITES regulations in arguing that it was 'extremely unlikely that the product came from Greenland' (IWC 54, Chair's Report).

Norway 'recalled' the 2001 suggestion that the IWC would be able to report to the next CITES meeting (in 2002) that the Revised Management Scheme (RMS) had been completed, describing the suggestion as having been 'too optimistic'. Norway then argued that the IWC's Report to the CITES in 2002 ought to 'give a true picture of the status of discussions on the RMS' (IWC 54, Chair's Report).

Norway, in its written opening statement, tried to lay down the gauntlet – declaring that:

> the IWC stands at a crossroads. We have reached a point where basic issues can no longer be dodged. Norway's position is crystal clear – we want the IWC to function in accordance with the principles and objectives laid down in the 1946 Convention.

Norway then suggested that it also wanted to 'see the IWC function in conformity with other contemporary, over-reaching international mechanisms pertaining to environment conservation and resource management', and that special note should be taken in this regard of the 1992 Convention on Biological Diversity and the 1982 UN Law of the Sea Convention, both of which, according to Norway, 'have served to further entrench the twin principles of conservation and sustainable use of Nature's resources'.[23] Norway concluded by warning that it saw a 'grave danger that undermining the principle of sustainability in the IWC will serve to undermine that principle within the whole structure of international cooperation on environment conservation and resource management'. According to the statement, Norway 'sees the developments in the IWC as a test case in defending and consolidating these basic principles', and that this is why the country has 'chosen to continue to work within the IWC with these considerations in mind, – still hopeful and optimistic that the IWC will decide to join this effort' (IWC 54, Opening Statements).

Kenya, in its (written) opening statement, stated that its 'policy on non-consumptive use of wildlife is well known'. Kenya then explained that taking this position 'has significantly boosted our social and economic development over the years, as Kenya's economy is largely dependent upon tourism, which is the leading foreign exchange earner and the only developing sector over the last three years' (IWC 54, Opening Statements). One thing that is noticeable about Kenya's stance is that it is consistent – the country's policy on whaling is echoed in its policy on the use of elephants and indeed most other wildlife. Kenya, however, no longer attends the IWC as it owes far too much in unpaid dues.

The fifth Special Meeting was held in October 2002, in Cambridge. The meeting was held at the request of the US and Russia, in order to repeat their request for a Schedule amendment to allow for the granting of an aboriginal subsistence take of bowhead whales from the Bering-Chukchi-Beaufort Seas stock – the aboriginal quota having been lost at IWC 54 in May 2002.[24]

Again, there was controversy over Iceland which had deposited an instrument of adherence with the US (as depositary government). Again, the reservation to the moratorium was an integral part of the instrument; however, Iceland had made a concession by committing not to authorise whaling 'for commercial purposes by Icelandic vessels before 2006'. The Chair decided that, as in previous years, it should first be decided whether or not the Commission has the competence to decide the issue, before, second, deciding whether or not to allow adherence with a reservation. After much debate, and several votes, Iceland was permitted to rejoin with a reservation (IWC Special Meeting 2002, Chair's Report).

The Special Meeting saw consideration both of a Schedule amendment in respect of the aboriginal subsistence hunts of the Bering-Chukchi-Beaufort bowhead whale

23 Despite this, when I asked Rodrigues Eusebio and Johannsen in 2007 how Norway might feel about closer links between the CBD and the IWC, or somehow bringing whaling under the auspices of the CBD, they reacted with alarm and clearly indicated a preference for stronger regional – instead of global – management.

24 Space has not permitted discussion of this, but at IWC 54 Japan led a number of parties in expressing concern over the scientific validity of the US–Russian quota and such parties successfully voted against renewal of the quota.

stock and of a proposal on Japanese coastal whaling. In an introductory statement, the commissioner for Russia put the position in emotional terms, arguing that:

> in order to save time the Russian delegation requests that this meeting should not discuss the US-Russian proposal but accept it on the basis of consensus, and that distinguished members of delegations take the floor only if they do not agree to the possibility of a consensus on the question of whether people should eat or starve.
>
> (IWC Special Meeting 2002, Chair's Report)

Russia and the US then proposed that a five-year quota of up to 280 for bowhead whales from the Bering-Chukchi-Beaufort Seas stock be approved, until and including the year 2007, with up to 67 strikes each year, except that up to 15 strikes might be carried over each year. The proposal also suggested that the provision be reviewed annually by the Commission, in light of the advice of the Scientific Committee. In debate, Japan indicated that it 'recognised the commitment of the US to review the provision on an annual basis and to abide by the advice of the Scientific Committee', these features being what distinguished the proposal from the proposal which had been rejected at IWC 54 in Shimonoseki. Japan indicated that while it appreciated the US' response, it did not feel that all of the questions Japan had raised at Shimonoseki had been addressed. Nevertheless, Japan, although not able to join the proposed amendment, would not block consensus on the amendment. Norway also indicated that it would 'support the consensus adoption of the proposed Schedule amendment', while recalling its stance that 'the Convention does not recognise the concept of aboriginal subsistence whaling' and that Norway considers the Commission's practice of 'distinguishing between commercial and subsistence whaling' to be 'artificial, illogical and morally wrong'. The proposed Schedule amendment was eventually adopted by consensus (IWC Special Meeting 2002, Chair's Report).

Japan then put forward a draft resolution requesting an interim relief allocation of 50 minke whales for its four community-based whaling communities, which request, according to Japan, 'had been denied for the last 15 years'. The draft resolution would clearly not, by itself, have seen the allocation of a minke whale quota to the four Japanese communities – but as a symbolic step toward such allocation it would have had huge importance.

Japan then advised that it 'considered that it had made a big effort to break the impasse on this issue', noted the delicate balance within which the work had been done and requested that the resolution be adopted by consensus. Countries which supported Japan's proposal included Antigua and Barbuda, Benin, China, Denmark, Guinea, Iceland, Morocco, Norway, Palau, Russia, the Solomon Islands, St Kitts and Nevis, St Lucia and, in a major turnaround, the US. The US advised that it was able to support the resolution as two of its major concerns had been addressed: that Japan satisfy the provisions of paragraph 10(e) of the Schedule, and that the Scientific Committee must advise that the 'catch from this stock of minke whales is sustainable' (the latter being a sub-paragraph added to Japan's draft Resolution for the first time). The US then advised that it considered Japan's draft resolution to show 'a marked change and a positive movement towards presenting a proposal acceptable to a broad range of the Commission' (IWC Special Meeting 2002, Chair's Report). One must be somewhat cynical, of course. While countries continued to phrase their comments to argue that they remained consistent with their usual positions, it was very clear that deals had been concluded behind the scenes.[25]

Countries like Australia, Austria, Brazil, Germany, Italy, Mexico, New Zealand, Sweden, Switzerland and the UK, however, advised that they were not able to support the draft resolution, considering that the proposal was inconsistent with paragraph 10(e), would establish (*de facto*) a new category of whaling and that it prejudged decisions made at IWC 55. The UK added that it considered the 'only quota consistent with paragraph 10(e)' to be zero. New Zealand considered that the commercial aspects of Japanese coastal whaling, which had been objected to previously, had not been addressed by Japan's efforts. Ireland suggested that it would not agree to change paragraph 10(e) 'before a satisfactory RMS' had been adopted. Antigua and Barbuda, on the other hand, suggested that it was 'remiss of the Commission not to deal fairly with Japan's Resolution', and that it believed that 'double standards were again being employed'. The Solomon Islands agreed. On being put to the vote, however, there were 16 votes for, 19 against and two abstentions (IWC Special Meeting 2002, Chair's Report).[26] Probably the most interesting aspect of this vote is that the US voted in favour – after more than a decade of firm opposition to such a proposal. The observer is entitled to raise at least one eyebrow.

That Russia and the US were denied their quota in Shimoseki does appear at first glance to have been very clever of the Japanese delegation in that it cunningly outmanoeuvred the US and, by forcing the US later that year to call another meeting, exposed US hypocrisy. Scientific uncertainty over bowhead numbers was the reason given. However, by proposing a 'logrolling' compromise amended proposal which would have given Japan a STCW (Small Type Coastal Whaling) quota for itself and – at the Special Meeting – not voting against the US quota, Japan also exposed its own hypocrisy. Neither side comes out of the episode with much credit (or credibility) at all (Couzens, 2012).

Komatsu and Misaki suggest that 'twenty years have passed since whales were given protection from commercial whaling. In this time, many species of whales have increased to the point where the balance of the ocean ecosystem has become skewed'. We must now, they continue,

25 I was advised by a pro-whaling delegate that the 'price' of Russia and the US obtaining the Aboriginal quota was Iceland's readherence. I was not given specific details on this.

26 The countries which voted in favour were Antigua and Barbuda, Benin, China, Denmark, Dominica, Grenada, the Republic of Guinea, Iceland, Japan, Norway, Palau, Russia, St Kitts and Nevis, St Lucia, the Solomon Islands and the US. Voting against were Australia, Austria, Brazil, Chile, Finland, France, Germany, Italy, Mexico, Monaco, the Netherlands, New Zealand, Peru, Portugal, San Marino, Spain, Sweden, Switzerland and the UK. Ireland and Korea abstained.

> call for the protection based on the sustainable use of balanced resource alignments in the ecosystem. Japan will seize the opportunity to call for the need of sustainable utilization of whale resources based on this principle at the IWC in Berlin in June, 2003.
>
> (Komatsu and Misaki, 2003: 8)

This may be the intention with which Japan began the year 2003, but its plans were to be blown out of the water.

Nicaragua made its first ever appearance at the IWC in 2003 and gave a (verbal) opening statement in which it 'noted' that the management of whales is 'inter-related with, and important to' fisheries management, and therefore that it is 'vital that developing nations dependent upon fishery resources participate in the work of the IWC'. Nicaragua suggested that in recent years attention has been diverted away from whale management and toward 'issues better carried out by others', and specifically suggested that 'trade in whale products is the function of CITES' (IWC 55, Chair's Report).

Mexico's draft resolution, entitled the 'Berlin Initiative on strengthening the conservation agenda of the IWC', proposed, inter alia, the establishment of 'a Conservation Committee of the IWC, composed of all Contracting Parties', with the Commission entrusting to this committee the power, essentially, to prepare and recommend, on an ongoing basis, a conservation agenda to the Commission. Mexico argued that 'there is a conservation agenda to be discussed and examined within the IWC'. The Initiative envisaged, per Mexico, the creation by the IWC of a dedicated conservation committee designed to investigate and address various threats to whales, such as over-fishing, climate change, toxic contamination, undersea noise, shipping and accidental catches. Mexico argued that the draft resolution was not, as suggested by Japan prior to the meeting, an 'attempt to change the fundamental purpose of the whaling treaty by introducing a strategy to end all sustainable use of whale resource for food'; but rather that the resolution was being proposed 'simply to provide the institutional structure needed to take account of conservation issues within the IWC'. Mexico noted further that the draft resolution was not an attempt to change the ICRW or to override the Scientific Committee. To an extent, those who argued in support of adoption of the resolution argued for linkages and an 'ecosystem approach' to management. Sweden, for instance, noted that 'conservation issues are not just linked to questions of sustainable use, since various forms of degradation and fisheries practices are threatening the world's whale populations'. On an interesting note, Monaco 'pleaded for consistency in countries' attitudes to conservation', suggesting that it 'did not see the sense of contracting governments voting for conservation in other fora and against it at IWC' (IWC 55, Chair's Report).

One important point to note here is that the anti-whaling countries shifted the terms of the debate somewhat by 'regaining control', so to speak, of the word 'conservation'. I will, however, continue generally to use the word in the sense of sustainable use, as opposed to 'preservation'.

On the other side of the debate, Denmark queried the legality of the draft resolution on procedural grounds, as it would be using a (majority vote) resolution to amend the Schedule; the country then suggested that the priority should be to complete the RMS. Iceland 'noted' that it is a supporter of 'whale conservation' and believed in sustainable whaling; but that it felt the draft resolution to be 'hijacking the terms of the Convention by selective quotation from its preamble' and that it would have the effect of drawing 'attention away from work on the RMS' and would 'increase polarisation in the IWC'. Norway, likewise, expressed the belief that the 'Berlin Initiative' would 'aggravate' the present imbalance in the IWC, in terms of which the 'orderly development of the whaling industry' was given insufficient weight, and that a 'radical and lasting change in the character of the IWC' would be created. Norway argued that the proper way to achieve such a result would be 'a diplomatic conference to renegotiate the Convention', and that this course, which it recognised not to be 'feasible', would be 'circumvented' by the Initiative. China argued that it 'did not see the need for a Conservation Committee', as the 1982 moratorium was itself a 'significant conservation measure'. Japan stated that it 'believed the draft resolution to be contrary to the primary objective' of the ICRW; and, noting that 'about half of the Commission appeared strongly opposed to the draft resolution', questioned how it might be able to 'function properly'. Of interest is the suggestion, by Russia, that 'whale conservation could be given high priority under the framework of the Convention on Biological Diversity (CBD)' (IWC 55, Chair's Report).[27]

Andresen writes that as changing the purpose of the ICRW would demand the consent of all parties, it is clearly not attainable (Andresen, 2001). However, in 2003 it seems that an attempt *was* made to change at least certain focal aspects of the IWC's role with the creation of the Conservation Committee. Reactions to the vote from both camps were equally opposed. According to Australia's then Environment Minister, David Kemp, the Initiative 'recognised' that the 'primary objective of the IWC is to conserve whale populations for the benefit of all humankind and for future generations' (quoted in Pohl, 2003). In contrast, according to Joji Morishita, speaking for the Japanese delegation, the 'Initiative is a Trojan horse. It looks harmless from the outside, but is actually destroying the IWC' (quoted in Pohl, 2003). According to Stefan Asmundsson, speaking for Iceland, what was witnessed was 'the concept of conservation being hijacked by protectionists. Conservation is a means to enable sustainable whaling. This proposal was not about conservation but protectionism and animal rights' (*ENS News*, 2003). Clearly, the battle is over preservation and conservation and the meanings of these words – and over different philosophies of natural resource use.

27 On this, however, Mexico responded that it did not believe that this would be an 'appropriate alternative since the CBD does not have competency over cetaceans, unlike the IWC', and that not all IWC members are party to the CBD (IWC 55, Chair's Report). (This last point was probably not well-made as the CBD has a high adherence rate with 193 parties by 2013, the overwhelming majority of which had ratified by the end of the 1990s.)

In its (written) opening statement, Australia had suggested that it was 'with a sense of history' that it welcomed the Berlin Initiative. Australia suggested that the Initiative:

> is a formal recognition of the reality that our Commission has evolved. It has evolved from an organisation that primarily attempted to manage the whaling industry, to an organisation that focuses its expertise on how best to conserve the marine biodiversity that this industry massively over-exploited.

Norway, on the other hand, in its (written) opening statement, argued that the:

> IWC was *not* established in order to obstruct, strangulate and ultimately liquidate the whaling industry. It was *not* established in order to undermine the principles of conservation and sustainable use. Nevertheless, the Commission has turned into an instrument for activists seeking to *prohibit* whaling!
>
> (IWC 55, Opening Statements)

It would seem, however, with hindsight, that the Berlin Initiative has not had much impact in the subsequent years – and in fact has not changed the IWC, or not significantly. At the same time, though, the divide between the protagonists in the IWC can seldom have appeared wider; to this day Japan does not attend Conservation Committee meetings.

The UK raised again the issue of the whale meat bought in Japan, but which had been labelled as from Greenland and Russia. The governments of Denmark, Japan and Russia had investigated, with the samples being divided for two different sets of DNA analysis. The analysis by Japan had shown that the product labelled as coming from Greenland was in fact Antarctic minke whale and the product labelled as coming from the Russian Federation was in fact Dall's porpoise. While stating that it did not doubt Japan's analysis, the UK wished that Japan would arrange CITES permits for the second set of samples, which were stored at the US embassy in Tokyo, to be analysed outside Japan. Japan indicated that it had 'no intention of issuing CITES permits' and the UK then indicated that it 'would try to arrange for corroborative analysis to be done in Japan'. Norway and Russia both 'noted' that they considered that trade issues are not within the IWC's competence, with Russia stating that it was 'completely satisfied with Japan's report'. The UK stated that it 'did not agree with this view on competency' and it believed that such issues were of crucial value to any RMS (IWC 55, Chair's Report). This incident does show rather ironic use by Japan of the provisions of CITES in order to refuse to cooperate with the UK's demands. The usual position sees Japan arguing that CITES restrictions on trade in whale meat ought to be relaxed and the UK arguing that they should be upheld. As is common in international law and politics, international law appears here to have been used as a political tool.

Australia discussed an opening statement put forward by the International Union for Conservation of Nature (IUCN), which Australia argued to be inappropriate as it 'provided highly specific and tendentious advice to the Commission on how it should manage progress on the RMS'. Australia argued that it was 'unusual' for the Secretariat[28] of one organisation to give such advice to another. This could be seen as somewhat ironic, given that Australia has supported the IWC advising CITES. Norway advised that the IUCN often does provide such advice, particularly to CITES; although Norway expressed a preference for the view of Australia (IWC 55, Chair's Report). Norway probably was seeing things more clearly here than Australia – Norway believing that the IWC should not dictate to CITES.

In October 2003, Nobutoshi Akao, of the ASEAN-Japan Center, argued that despite Japan's veiled threats that it might leave the IWC, 'Japan should neither secede from the International Whaling Commission nor suspend its financial contribution. Rather it should remain in the IWC and continue IWC-reforming efforts'. Japan's seceding from the IWC, Akao argued, 'would mean that Japan would lose international opportunities to make its justifiable case and would be isolated internationally' (*JWA News*, 2003). This could be a useful clue as to why Japan remains in the IWC. Japan has perhaps learned from observing Iceland, which withdrew from the IWC, hoping to be able to exercise more authority over its whaling affairs from outside, but found this difficult and has since returned. In addition, Japan has put too substantial an effort over the years into reforming the IWC to withdraw easily.

Friedheim asks what kind of threat could be made that would be meaningful to the anti-whaling coalition, and suggests that in respect of withdrawal from the IWC 'unrestricted whaling is not a credible threat, but the creation of a rival organization might be' (2001c: 223). Japan appears, however, to be trying still to influence the IWC from inside. This may explain, at least partly, recent increases in IWC membership. Oddly, this has led to some ironies. As Japan has played the numbers game, so have the anti-whaling states. The anti-whaling states even appear to have, for the present at least, won the race to acquire majority support. A second problem is that as pro-whaling states have sought to keep the IWC separate from other international environmental conventions, the introduction of so many new states has made it less tenable to argue that the IWC is not inherently linked to various environmental conventions.

In advance of IWC 56 in 2004, Japan warned again that it might leave the commission; but despite the occasional bit of sabre-rattling, it clearly appears that Japan does not *want* to pull out of the IWC. This can arguably be seen as a hopeful sign for international law and cooperation. Denmark, in its (written) opening statement to IWC 56, suggested that the 'IWC has come to a critical point in its recent history. It is now time for the IWC to return to its responsibility as an organisation for *both* the management *and* the conservation of the great whales'. Denmark argued, therefore, that the RMS, including 'a robust and cost-effective inspection and control scheme for future commercial whaling', be 'approved and fully implemented' as soon as is possible (IWC 56, Opening Statements). It is interesting to note that Denmark chose to split 'management' and 'conservation',

28 It is unclear what Australia meant by 'Secretariat', referring to the IUCN.

as though the former means use and the latter means preservation. It is also interesting to see Denmark suggesting a dual role for the IWC – Denmark being both a member of the anti-whaling European Union and at the same time (through Greenland and the Faroes) a whaling state.

Japan, in its (written) opening statement, also suggested that the meeting was a critical one. Japan stated that both the general public and parliamentarians in Japan have come to 'the end of their patience on this matter and demand that the Japanese Government consider all options to resume sustainable whaling if the IWC has not implemented a reasonable RMS' by IWC 57. Japan then declared that 'paragraph 10(e) of the Schedule is no longer in force due to its actual wording ("and by 1990 at the latest")' and due to 'the unnecessarily long delays in implementing the RMS for the resumption of commercial whaling on a sustainable basis'. Norway agreed, asserting in its (written) opening statement that 'the moratorium is no longer binding on the parties to the convention, because of the failure of the Commission to honour the provisions set out in paragraph 10(e)'. 'Furthermore', argued Denmark, 'the parties to the Convention neither empowered the IWC to declare a moratorium on whaling nor the authority to establish sanctuaries without basing the decision on proper scientific evidence and recommendations' (IWC 56, Opening Statements).

Germany, in its (written) opening statement, used much the same language as had Denmark, suggesting that 'Germany recognises both mandates laid down in the IWC Convention, that is to say the conservation *and* management of whale stocks'. 'While the Conservation Committee', argued Germany, 'should devote itself to whale conservation, it should on no account prevent the management of stocks'. India, in its (written) opening statement, argued that, apart from 'illegal harvesting', factors like environmental pollution, climatic changes and bycatch, mean that 'the challenges faced by whales are much more serious than ever before', and that, in consequence, 'there is now a greater need for international cooperation for protecting whales than ever before in the history of the IWC'. India stated that it 'would like to remind all that the whale is not one giant meat ball' and that the 'right of the whale in its natural environment to roam unhindered and unthreatened in the blue waters that constitute two-third of our world must be recognized without any qualification' (IWC 56, Opening Statements). It is interesting to note that India is a country which can probably claim to act consistently in respect of both the elephant and the whale – being opposed to the trade and exploitation of both species.

A number of countries – Japan and Denmark especially – had gone into the meeting declaring that it would be a 'crucial' one. With hindsight, it probably was not as significant as it was billed by several countries to be; although change did occur in subtle ways – such as in the joinder of four new apparently pro-whaling countries (Côte d'Ivoire, Mauritania, Suriname and Tuvalu) which somewhat narrowed the gap between the polarised voting blocs.

At IWC 57, in 2005, nine new member nations joined the IWC. At the end of the meeting, it was reported that the meeting had ended with the usual impasse (Black, 2005). *JWA News* reported, however, that the meeting had been characterised by the fact that the votes of pro-use countries and of anti-whaling members had been almost equalled, with the total number of IWC members increasing from 57 to 66. 'IWC watchers,' said this Report optimistically, 'observe that the moves of additional new members and existing middle-of-the-road members will hold the key to the direction of future IWC meeting' (*JWA News*, 2005a).

3.7 A victory for the pro-whaling side

The year 2006 appeared to see both an escalation of whaling efforts, to mounting alarm in the anti-whaling camp, and of protests against these. In April, the averment was made that 'while the world has been looking the other way', the Japanese have 'spent nearly a decade and many millions of dollars building up a voting majority in the IWC, by buying the votes of small member states with substantial foreign aid packages'. Concern was then expressed that Japan, with numerical control, would be able to take significant steps toward having the moratorium lifted; such steps included the introduction of secret ballots (*The Independent*, 2006). The problem with this argument, of course, is that – if followed to its logical conclusion – it would imply that Japan is correct to argue that a significant number of countries really *do* wish to vote other than as they do, for whatever reason, and to support a resumption of commercial whaling.

The IWC meeting was set to run from mid-May to mid-June 2006, in St Kitts and Nevis. The attitude of the host country was evident from a report in April, where a Senior Fisheries Officer in the St Kitts and Nevis Ministry of Agriculture was quoted as saying that 'the stance of some countries and agencies with regard to the issue of whaling is based on emotion and not science' (*caribbeannews*, 2006). In May 2006, it was reported in *National Geographic* that Paul Watson of Sea Shepherd had argued that there was 'no debate', as the Japanese whalers are 'acting commercially under the auspices of "bogus research" and therefore are in violation of the 1986 moratorium' and that the whalers 'are also in clear conflict with the Convention on International Trade in Endangered Species (CITES)' (Heller, 2006). Obviously, there are at least two difficulties with this view. That the research conducted by Japan is not genuine is an interesting argument, but at this stage remains only a contention. It is clear-cut that genuine research would be protected (and that there would then be an obligation to utilise the whale meat); it is also implied that 'bogus' research would not be protected under the ICRW. Between these two extremes, however, the water is murky – how is one to determine whether the research is genuine or not? Ultimately, the only way to determine this would be to have the IWC rule – but the IWC's ruling would depend on a political majority which might not be genuinely voting as it believed. As to the point about CITES, this is clearly incorrect if what is meant is whale meat being traded across international borders. If the alleged infringement is of 'introduction from the sea' then the point is better made, but might be covered by CITES reservations held by Japan. It is interesting, however, to see an opponent of whaling claiming infringement of both CITES and the ICRW.

Something of the frenzy which began to grip the Western media can be seen in a report from the normally fairly sober UK newspaper *The Guardian*, also in late May, in which it was reported that about 35 pro-whaling countries were expected to gain control of the 66-member IWC; that 'in a triumph of intensive lobbying led by Japan, they are likely to dismantle the laws that protect whales and prepare for the eventual full resumption of commercial whaling'; and that:

> to the horror of more than 150 large conservation organisations worldwide, the anti-whaling countries led by Britain, New Zealand, Australia and the US, say there is little that can now be done diplomatically to prevent the takeover by pro-whalers at the IWC meeting.
>
> (Vidal and McCurry, 2006)

In June, importantly when considering linkages between the IWC and CITES, Norway's commissioner, Karsten Klepsvik, was quoted as saying that 'a simple majority at the IWC would be enough to request CITES to lift a ban on trade in minke whales' (Christie, 2006). This is a significant comment – being a direct suggestion from an important IWC commissioner that CITES might be used to circumvent the IWC. Even further, the suggestion implies that CITES might be used to make the IWC irrelevant. As a possible consequence of the obtaining of a numerical majority, this shows how important the linkage is. Just a year later, at IWC 59, Karsten Klepsvik confirmed to me directly, albeit at tea and not in a formal interview, that Norway does see important similarities and linkages between the two treaties (Klepsvik, pers. comm., 2007).

Accusations of bribery flew. However, the anti-whaling countries do have a difficult task convincing the impartial observer that they are not themselves guilty of hypocrisy. While the charge is made – by anti-whaling states and NGOs – that the pro-whaling states have 'bribed' smaller countries, bringing in small countries to support your position is not a practice only of the pro-whaling states. It was, for instance, reported in June that Israel had joined. The sceptical observer might ask whether Israel joined because it wanted to, in the belief that it had a meaningful role to play, or because its political interests are best served simply by supporting the US with its vote – and that this is the only 'contribution' which it will make in the IWC. In regard to Israel's joining, it was reported in June that Israel's new government had not yet formulated an official policy on the issue, but that Israel had joined after a direct plea from the US to 'help defend the moratorium on commercial whaling' (*ENS News*, 2006b). This perhaps makes Israel the single country that has joined the IWC without claiming to be doing it for its own reasons.

The Japanese themselves played down the possibility of their gaining a numerical majority. In an interview in mid-June, Morishita said that he could not officially confirm numbers. In his words,

> there have been some additional members coming, but like last year, some members might not show up and many of the sustainable use countries are from poor, developing countries. They may not be able to pay a contribution,

> then they don't have a voting right. So until the first day comes out, we don't know exactly what the voting balance is.
>
> (Clarke, 2006)

In January, the Sea Shepherd Conservation Society had alleged that the South African authorities were aiding the Japanese after the organisation's ship, the *Farley Mowat*, was detained in Cape Town. It appears that the South African Maritime Authority (SAMA) detained the ship because it did not have a maritime safety certificate – Director Paul Watson's counter argument was that the ship was registered in Canada as a yacht, and therefore did not legally require a safety certificate (Sea Shepherd, 2006). It is a strange accusation to make, that South Africa was 'in cahoots' with Japan on this one – given that South Africa's stance appears firmly anti-whaling. Unfortunately, Sea Shepherd gave no details on its website to support the charge, nor even referred to the existence of documentary proof. Of course, it *is* possible that South Africa could choose to cooperate with Japan when it has nothing to lose by doing so. I questioned Paul Watson on this in Anchorage, Alaska, in May 2007, and Watson's answer was that he had no 'actual' proof, but that he did have strong suspicion based particularly on the fact that the South African authorities had had an arrest warrant already prepared before the Sea Shepherd ship arrived in Cape Town. This, he concluded, meant that somebody must have given the South African authorities essential details in advance. When I put to Watson that he had suggested, at least several times, that the governments of Japan and South Africa had colluded to prevent the Sea Shepherd ship from leaving Cape Town, and asked him whether he had any evidence of such collusion, his reply was that it is:

> hard to really get hard evidence, but when we arrived in Cape Town they had already issued the detainment order – the day before we had arrived – and then they came on board and said they wouldn't recognise us as a yacht.

Watson's take on the matter was that although SAMA claimed that their actions were 'routine', he, Watson, had been 'doing this for a long time', and 'recognised harassment when he saw it' (Watson, pers. comm., 2007).

Oosthuizen told me that he had himself asked SAMA the same question and had been told that it (collusion) was not the case at all, but that it had been purely a safety resolution (Oosthuizen, pers. comm., 2006). Horst Kleinschmidt felt much the same way, saying that he felt that if there had been any pressure behind the scenes it would have come from Canada – and that allegations of direct collusion between Japan and South Africa were far-fetched. He felt that the South African officials had 'boxed themselves into a corner' – based on Canada having advised them that Canadian legislation had changed and the definition of a 'sailing vessel' having changed (Kleinschmidt, pers. comm., 2006). Butterworth felt that the most that could have happened was that Japan might have 'brought certain things to diplomatic attention', but that ultimately the country was simply 'not that unsubtle' (Butterworth, pers. comm., 2006).

I spent so much time on this point because proof of collusion between Japan and South Africa in a whaling context would be of huge significance for the arguments being made in this book. Ultimately, however, it seems unlikely to me that there *was* such collusion – for the very simple reason that there would have been little point in it. Sea Shepherd had concluded its campaign against the Japanese scientific whaling effort for 2005/6, and was on its way *from*, not *to*, Antarctica when the *Farley Mowat* docked in Cape Town. As such, deliberate harassment would have been done purely for spite, and this seems unlikely.

By mid-June, information was being released from all sides as the media frenzy mounted. It all seemed rather far from the IWC meeting of only three years previously, when the anti-whaling side was trumpeting the Berlin Initiative and claiming to have turned the IWC around and to have made it a conservation ('preservation') treaty. Reports from the opening days of the IWC meeting, however, suggested that the environmentalist critics had indeed been alarmist in the Western media. It was reported that at the outset Japan had lost a vote to stop the IWC discussing small cetaceans. According to one report, the '32-30 vote against a proposal to remove so-called small cetaceans from the IWC's agenda at a gathering on the Caribbean island state was seen as a bellwether of the balance of power' (*MSNBC.com*, 2006). After the vote, Morishita was reported to have downplayed his country's interest in securing a majority, reportedly saying that 'we aren't so obsessed with the majority' and that 'we want to make sure we can get some countries to accept the fact that we can at some point move to sustainable commercial hunting of whales' (*Jamaica Observer*, 2006).

The pro-whaling side then lost a second vote. Japan had proposed bringing in secret ballots, arguably so that small Pacific and Caribbean countries could back its pro-whaling stance without criticism from environmentalists, but this proposal was defeated by 30 votes to 33, with one abstention. What appears to have happened is that the Solomon Islands, which was expected to side with Japan, had abstained on this vote and that Belize had unexpectedly voted against both initiatives. Of interest was the behaviour of the Solomon Islands, in regard to the vote on secret ballots. It was reported that in 2005 the country had made a commitment to Australia that they would support anti-whaling nations, but had then voted with Japan. In 2006, however, the country abstained, thereby representing some return for Australia, which had lobbied extensively in the weeks leading up to the vote (*ABCOnline*, 2006).

Japan then put forward a proposal, regularly put forward, which would allow Japanese coastal communities an interim quota. The vote ran 30 for and 31 against, with four abstentions. The abstentions were China, Kiribati, the Solomon Islands and Korea – at least some of whom might have been expected to vote in favour. According to Sea Shepherd, the vote prompted Morishita, the Japanese delegation's spokesman, to say, 'we are glad this is not a secret vote. Japan will remember which countries supported this proposal and which countries said no' (Sea Shepherd, 2006). This would be an important comment, if true. One cannot take Sea Shepherd as giving reliable comment, but other sources have given the same wording (for instance, *ENS News*, 2006a). It is a more clumsy threat than might have otherwise been expected from the urbane Morishita, but might show something of the emotion – and the frustration – felt in the immediate aftermath of losing such a vote so narrowly. The fourth vote was a Japanese proposal to eliminate the Southern Ocean Whale Sanctuary, which would have required a 75 per cent majority vote and was therefore of symbolic value. The vote in the end was 28 for, 33 against, with four abstentions.

In a real surprise, however, the next vote (on the proposed 'St Kitts and Nevis Declaration') was successful with 33 for, 32 against and one abstention.[29] Sea Shepherd acknowledged that the resolution 'was put forward by the host nation of St Kitts and Nevis' but clearly believed that Japan was responsible (Sea Shepherd, 2006). In fact, it probably was not – at least not according to Rune Frøvik who told me otherwise (pers. comm., 2007). Amongst various reactions to the vote, a spokesman for the Japanese delegation was described as saying that the vote was 'a historic moment', was 'the first serious setback for those against whaling in years' and that it was now 'only a matter of time before the commercial ban is overturned' (Raney, 2006).

It was reported that 'Japan was jubilant' and that 'a foreign ministry spokesman in Tokyo hailed the passage of a resolution critical of the ban' as 'a significant step forward' (McCurry, 2006). Another report put the sense of victory far more strongly, suggesting that the 'environmental movement' had:

> suffered one of its greatest reverses when pro-whaling countries, led by Japan, gained control of the IWC and immediately began undermining the 20-year-old moratorium. In a stunning diplomatic coup, Japan and its allies, including Norway and Iceland, won a voting majority in the IWC for the first time.

Denmark, the same report continued, was one of the 33 countries to vote in favour, apparently declaring that the 'IWC has failed to meet its obligations' and that it needed to be 'normalised' (McNeill, 2006). That Denmark broke ranks with fellow EU members and voted in favour was a surprise.

It was reported in the Norwegian media that 'fans of Norwegian whaling were triumphant Monday after the IWC approved a resolution that may finally make Norway's controversial annual whale hunt legal' (Berglund, 2006). This is a somewhat odd description to find in a Norwegian media article as it implies that the Norwegian hunt is not currently legal. Certainly, members of the Norwegian delegation would take exception to any charge that the country is not acting legally in taking whales under a valid reservation. The report continued, arguing that:

29 The St Kitts and Nevis Declaration (Resolution 2006-1) said, in essence, that the ICRW was a sustainable use instrument; that the moratorium was no longer necessary; that protecting all whale species regardless of stock status was contrary to the object and purpose of the ICRW; and that the IWC should be 'normalised'.

> the anti-whaling sentiments within the IWC appear to be shifting. Norway, Japan and other whaling nations like Iceland and Russia gathered enough support from Caribbean and African nations to push through a resolution that symbolically expresses an intention to ultimately allow commercial whaling once again.
>
> (Berglund, 2006)

It is certainly of interest that several reports have described Russia as being 'pro-whaling'. Although bolshy and self-interested, Russia has in recent years tended to favour the anti-whaling group. Were Russia to consider resuming commercial whaling, however, it would be a dramatic step.

Ultimately, the St Kitts and Nevis Declaration was merely an assertion won by majority vote. Its immediate consequences, though, were that it provided a firm statement as to the beliefs of the pro-whaling members, and, ironically, it was probably the best thing that had happened to anti-whaling NGOs in years as it allowed them legitimately to sound an alarm call.

3.8 The 'Future of the IWC' begins

IWC 59, in 2007, was held in Anchorage, Alaska. Throughout IWC 59 in Anchorage, there were meetings – bilaterals, small groups and larger coalitions. Throughout my first IWC meeting, I had the sense that things were always busy; that much was happening behind the scenes; and that innumerable deals, half-deals, promises and vague commitments were being made continually. Even where no actual deals were being struck, parties were all insatiably trying to find out what others' positions were and how others' views might change, no matter how slightly. There also was, in my opinion, a pervasive sense of something that approached paranoia – everybody present seemed to want to know about any possibility of there being any 'moves' to broker deals or switch stances.

My perception was that the main issue was indeed, as expected, aboriginal subsistence whaling, and that states were worried about the possibility of Japan from a distance orchestrating the blocking of the US-Russian bowhead and gray whale quotas, even though it was rumoured that Japan had apparently given private assurances that it had no such plans. States seemed to be worried about the potential for disruption that the aboriginal issue held – the events of Shimonoseki in 2002 being alluded to many times. Also, states were concerned about the potential for disruption caused by Greenland seeking to increase its quota, and to begin taking humpback whales. Further, states were very concerned that 'bundling' of issues and quotas might lead to the necessity of unwelcome compromises. Generally, consensus seemed to be that the matter of aboriginal subsistence whaling should be wrapped up as fast as possible, with the concern being that the longer it was kept open, the more Japan might try to create linkages. Further, it seemed generally agreed that there needed to be a resolution on CITES, which would meet immediately after the IWC, and that a strong message needed to be sent on the matter.[30]

Both Australia and New Zealand made emotional appeals from the floor to Japan to abandon its plan to introduce humpback whales to the list of whales taken for research purposes, with the New Zealand commissioner in particular suggesting that humpbacks are an iconic species – to New Zealanders perhaps even more so than are any other whales. Japan replied, in a conciliatory way, that more dialogue was needed and that they were open to this. The Commission discussed cooperation with other international organisations – concurring that such cooperation is 'an important part of its work, especially in the scientific arena' (IWC 59, 2007). Six new countries had joined the IWC, bringing the total membership to 76, of which 71 were present. By the end of the first day, the general feeling on the floor seemed to be that the meeting had been very relaxed; in stark contrast to other meetings in recent years (and especially to IWC 58).

On the second day, the Commission began considering the issue of aboriginal subsistence whaling and agreed on five-year renewal periods, as the usual term in recent years, for: bowhead whales to be taken by the indigenous peoples of Alaska and Chukotka; for gray whales to be taken by the indigenous Chukotkan people; and for humpback whales to be taken by the indigenous people of St Vincent and the Grenadines. Discussions on whales to be taken from Greenland continued. The catch limits, as requested, for the Alaskan, Chukotkan and St Vincent and the Grenadines catches, remained the same as for the period 2002–2007; the request from Greenland differed in that higher numbers of minke whales were requested, and additional species were requested (IWC 59, Chair's Report).

In debate before the decision, the US based its plea for a bowhead quota on the Scientific Committee Report, and on the contention that the 'uncertainties' of 2002 have 'been dealt with'. Most countries which spoke did so in support of the granting of the quota. Japan then stated that its policy on whaling 'is well-known – science-based and on sustainable utilisation'. Japan 'congratulated' the 'American, Japanese, Norwegian and other scientists in the Scientific Committee'. 'Based on our own policy', said Japan, 'and on the spirit of consensus' it would support the proposal; but then added the loaded remark that 'we only ask that the same criteria be applied to our coastal whaling later'. Iceland stated that it was 'happy' to note the 'general support for sustainable whaling', and added, with heavy sarcasm, that it 'will be interesting to see if we'll have similar support later – or several u-turns'. The proposal was adopted by consensus – to much applause, and a feeling of palpable relief among the countries comprising the Like-Minded Group. However,

30 I have been extremely circumspect on the attitudes of the Like-Minded Group, which met in private strategy sessions several times in order not to betray confidences. The practice has arisen in recent years of referring to the majority of the anti-whaling parties as the 'Like-Minded Group' and the majority of the pro-whaling parties as the 'Sustainable Use Group' – the parties themselves using these terms. Membership of such groups may be fluid at times, and the terms may not accurately describe parties' positions. South Africa, for instance, is in the Like-Minded Group; but this should not be taken to mean that South Africa's philosophy is not that of sustainable use.

to me it seemed as though Japan had – despite apparently promising behind the scenes not to link the bowhead quota with other issues – cleverly created just such linkage.

Parties spoke for and against the Greenland proposal – particularly on the issue of the request to take up to ten humpback whales annually. Iceland (which suggested not taking an 'approach of the survival of the cutest'), Japan and Norway all spoke firmly in support of the proposal. Norway referred, sarcastically, to serious concerns having been raised over Alaskan bowhead whales some years earlier, and pointed out that in that case an interim quota had been given. Various countries then argued against the proposal. There was a short discussion of the word 'aboriginal'. St Lucia described the word as a 'derogatory term'; and St Kitts and Nevis stated that the word was a colonial term thrust upon them and suggested putting a working group together to 'solve this problem'. The matter was left open. Progress made on the Revised Management Procedure was discussed, but the agenda item kept open – largely because at IWC 58 the Commission had accepted that an impasse had been reached on RMS discussions, and no formal activity was identified for work prior to IWC 59.

The Commission then considered an Agenda item titled the 'Future of the IWC'. Three non-IWC meetings which had taken place in the previous year were referred to – one held in Tokyo, organised by the Government of Japan; one held in New York, organised by the Pew Foundation; and one held in Buenos Aires, organised by Latin American countries. The Commission recognised the 'difficulties in reaching consensus when views within the Commission are so polarised'; but many countries did note that 'there was positive overlap' in some of the suggestions made at the three meetings, and agreed that there was 'merit in pursuing intersessional work'. It was further agreed that there would be merit in pursuing this intersessional work 'through an intersessional meeting under the auspices of the IWC',which would be open to all contracting governments and observers (IWC 59, 2007).

The third day began with further discussion, which had started the previous day but not been completed, of the proposal – put forward by Argentina, Brazil and South Africa – to establish a South Atlantic whale sanctuary. Opinion was divided on the issue, and the proposal was put to a vote. With 39 votes in favour, 29 against and three abstentions, the proposal – which required a three-quarter majority – failed.

The Commission then considered matters related to socio-economic implications and small-type whaling, with Japan, as it had done in previous years, giving a presentation on the hardship allegedly being suffered by its four community-based whaling communities (Abashiri, Ayukawa, Taiji and Wadaura). Japan put forward a proposal that would allow a catch by these communities. After some discussion, however, it was clear that there was no consensus and the item was therefore left open. Japan's proposal was to amend the Schedule to allow for the taking of minke whales from the Okhotsk Sea/West Pacific stock for each of the years 2007–2011, with the meat and products to be used for local consumption exclusively (IWC 59, 2007). In discussion, Japan's representative, Morishita, pointed out that Japan had now for 20 years been requesting assistance for its four coastal whaling communities. He described the 'history of small-type whaling' as being a 'history of broken promises', and then pointed out that Japan had objected to the moratorium in 1982 but that the country had been put under 'strong pressure' from the US – in terms of the Packwood-Magnusson Amendment – 'as we were taking fish just off Alaska'. He then suggested that Japan had 'negotiated to keep the moratorium off for two years, if the US would let us keep our fish quotas' – and 'withdrew our objection on the understanding that the US would keep its promise' – but added that 'two years later we lost our fish quota'. This, he said, was the 'start of mistrust' and the 'first promise broken'. He then referred to the moratorium and the fact that the Schedule amendment states very clearly that there will be a review by 1990, but that it has been 'seventeen years and nothing has happened'. He called this 'another broken promise'. He then pointed out that in 1994 the RMP was adopted by consensus, but that the Commission has not acted, which he characterised as 'another broken promise'. He then pointed to the RMS, which had been under negotiation 'for thirteen or fourteen years' and that 'last year the Commission postponed discussion' – which he called 'another broken promise'.

Morishita then recalled that the Commission adopted resolutions on coastal whaling in 1993, 1995, 1996, 2000, 2001 and 2004 – all of which 'recognised the needs of the four communities and the effect of the moratorium' – but, sarcastically, suggested that in the IWC the word 'expeditiously' clearly 'has no meaning'. He referred again to a 'long, long history of broken promises'. He referred to 'commerciality' as one of the reasons given for distinguishing between small-type coastal whaling and aboriginal subsistence whaling, but pointed out that 'all activities are commercial' and that 'it doesn't make sense' that Alaskan whalers could sell products from bowhead whales. He agreed that Alaskan whaling 'should be fully supported', but that 'we should be treated the same'. He explained that Japan was not asking for a lifting of the moratorium, nor for a specific number of whales (as they were willing to negotiate on the number), but that Japan was willing to accept observers, and to 'put VMS on the small vessels'. 'I kept my promises', he said, 'and supported all aboriginal subsistence whaling proposals' – as 'respect for culture and way of life'.[31] Bringing his speech toward a conclusion, Morishita said that if the Japanese proposal was not supported, it would be 'sending a very bad message to us', and that 'we are approaching in a spirit of compromise and willingness to talk'. 'But', he said, 'we have been waiting twenty years', and, while Japan has 'some hope for the normalisation process', without seeing 'some tangible progress we will have to think of alternatives'.

The Commission considered the issue of scientific permit whaling. The Scientific Committee had, at an intersessional meeting earlier in 2007, carried out a review of the 18-year JARPA (Japan Whale Research Programme under Special Permit in the Antarctic) programme – after the completion of JARPA, Japan had begun a

31 This being a strong hint that, as discussed earlier, Japan does seem to have given assurances behind the scenes that it would not seek to block the renewal of the bowhead whale quota put forward by Russia and the US.

programme known as JARPA II. In 2006/7, 505 Antarctic minke whales and three fin whales had been caught under this programme, and under Japan's North Pacific programme a total of 195 common minke, 50 Bryde's, 100 sei and six sperm whales were caught in 2006. Iceland had taken 58 common minke whales as part of its programme (IWC 59, 2007). The Scientific Committee had apparently concurred with the conclusion of a review workshop – held in Tokyo in December 2006 – that 'the results of the JARPA programme, while not required for management under the RMP, have the potential to improve management of minke whales in the Southern Hemisphere in a number of ways'. The Commission then passed a resolution asking Japan 'to refrain from issuing a permit for JARPA II; or at least the lethal aspects thereof'. The voting ran 40 in favour, two against and one abstention, with 27 countries 'not participating' as they 'believed that the submission of the proposal was not conducive to building bridges within the Commission' (IWC 59, Chair's Summary Report).

The Commission then adopted a resolution on the non-lethal use of whales with 42 votes in favour, two against and two abstentions. Twenty countries did not participate on the ground that they felt that, given more time, a consensus resolution could have been developed. Brazil insisted that the resolution be put to the vote, which left me feeling that the *real* purpose behind it was that the Latin American group, which appeared to be flexing its muscles somewhat, wished to leave IWC 59 with a successfully sponsored resolution to its credit.

On the final day, discussion began with completion of the Agenda item on aboriginal subsistence catch limits for Greenland. The catch limits were eventually adopted by vote with 41 in favour, 11 against and 16 abstentions – the necessary three-quarters majority was attained. The original controversial request for humpback whales was dropped.

A resolution on the relationship between the IWC and CITES was passed with 37 votes in favour, four against and four abstentions – and 26 countries not participating in the vote, on the ground that they believed it was 'inappropriate'.[32]

32 Resolution 2007-4, 'Resolution on CITES', reads as follows:

> RECOGNISING that the [IWC] is the internationally competent organisation for the conservation and management of whale stocks; FURTHER RECOGNISING that the [CITES] passed Resolution Conference 11.4 (Rev. COP 12) which acknowledges the IWC as the major source of information on whale stocks around the world; NOTING that the IWC Scientific Committee continuously reviews the status of all whale stocks; NOTING that the moratorium on commercial whaling has been in effect since 1986, remains in effect and the reason for the moratorium remain valid; WELCOMING the continuing cooperation between CITES and the IWC on issues related to international trade in whale products, and urging all governments to continue to support IWC and CITES obligations with respect to this issue; FURTHER NOTING the existence of CITES Resolution Conference 11.4 (Rev. CoP 12) on the Conservation of cetaceans, trade in cetacean specimens and the relationship with the [IWC] which *inter alia* expresses concern that international trade in meat and other products of whales is lacking adequate international monitoring or control, recognises that the IWC is the major source of information on whale stocks around the world and recommends that the Parties to CITES agree not to issue any

In discussion on this issue, the UK – which had promoted the resolution – said that 'any increase in trade will put pressure on the moratorium', and that it was 'important this meeting sends a clear message to CITES'. Japan said that 'this resolution is about the relationship between CITES and IWC' and that 'we all know the issue is the moratorium'. CITES, said Japan, 'has been repeatedly asking the IWC to finish the RMP' and the 'reason is CITES wants to make its decision based on science' but 'last year we just postponed the RMS'. Japan then added the words 'it's effectively dead, unfortunately'. Japan then referred to Article 65 of UNCLOS and its mention of 'plural of organisations' and suggested that 'if IWC cannot do its job, other organisations can'. The country then stated that it would not participate in the vote, on the ground that it thought it 'wrong to tell other organisations'. Iceland, Mali, Russia[33] and St Kitts and Nevis spoke against the resolution (IWC 59, 2007). The UK declined to withdraw the proposed resolution, despite the requests of several countries, and pointed out that the 'proposals to review' that had been put to CITES CoP 14 'were from Japan and Iceland' and that the 'boot' was therefore 'on the wrong foot', and that the UK would press on to a vote.

The vote was then held and the resolution adopted with 37 in favour, four against (China, Dominica, Norway and Russia), four abstentions (Denmark, Mali, the Netherlands – perhaps because of its role as host of the upcoming CITES meeting – and Oman), one not present (South Africa) and 26 countries (including Iceland and Japan) 'not participating'.

Many strategies play out in international fora and it is extraordinarily difficult sometimes to determine why a state actor makes the decisions it does. In their

import or export permit, or certificate for introduction from the sea under CITES for primarily commercial purposes for any specimen of a species or stock protected from commercial whaling by the [ICRW]; NOW THEREFORE THE COMMISSION: AFFIRMS that the moratorium on commercial whaling remains in place and that the reasons for the moratorium are still relevant; EXPRESSES APPRECIATION that CITES recognises the IWC's Scientific Committee as the universally recognised international organisation with international expertise to review and evaluate the status of the world's whale stocks; REAFFIRMS the important role of CITES in supporting the IWC's management decisions with regard to the conservation of whale stocks and the importance of continued cooperation between CITES and IWC; REAFFIRMS the importance of continued cooperation between CITES and IWC with regard to the conservation of whale stocks through the regulation and management of international trade in whale products; CONSIDERS that the IWC has not yet completed the necessary measures to regulate commercial whaling; CONSIDERS that any weakening of existing restrictions on trade under CITES could have significant adverse effects on the moratorium on commercial whaling and increase threats to whales; REQUESTS Contracting Governments to respect the relationship between the two conventions and not to seek the transfer of cetacean species from CITES Appendix I. FURTHER REQUESTS the Secretariat to send a copy of this resolution to the CITES Secretariat.

33 Russia pointed out that it had a reservation to the moratorium, which it 'doesn't intend to use', on commercial whaling; and that it had voluntarily revoked its reservation in CITES. Russia then called for the resolution proposal to be 'rescinded in the interest of consensus'.

purest form, of course, decisions are reflected in the votes which a state casts at international meetings – but even votes are not always easy to understand. Technically, there are only three ways to vote: by voting 'yes', voting 'no' or abstaining. However, when a political protest seems desirable, or one of these choices might put a state into a difficult political position, another course might be followed. By way of political protest, in 2007 many pro-whaling states, knowing that they were numerically outnumbered and bound to lose any resolution voted on anyway, announced instead that they were 'not participating'. Sometimes, too, states might simply choose to avoid the issue, and when a delegate is absent this might even be seen as a form of 'voting'. Outside of IWC 59, I have been informed (off the record) of a state which is expected to vote in a certain direction, and usually does, but which has occasionally avoided the issue through the tactic of delaying payment of its annual subscriptions (and therefore having its vote suspended) until after the contentious vote has been held.

On the issue of Japan's proposed Schedule amendment to allow the taking of minke whales by its four community-based communities, Japan explained to the Commission that it had tried to include a new idea – that of requesting 'the Scientific Committee to develop a species specific method'. Japan asked for adoption of its proposal by consensus, but stated that it expected 'some opinions', that it had no intention of entering 'a lengthy discussion' and that after hearing opinions on the proposed resolution it would decide whether to withdraw the proposed Schedule amendment or not. After hearing opposed opinions from just a few countries (Costa Rica, New Zealand and the US), the Chair stated that there was no consensus and referred the matter back to Japan. Morishita, for Japan, then said that it was clear that small-type coastal whaling was not going to be accepted as aboriginal subsistence whaling, even though Japan felt that it was 'logical to look for a way out' as the IWC 'has previously recognised small-type coastal whaling'. He then explained that he had 'no illusions' and that 'pushing for a vote will probably increase division'; he would therefore not ask for a vote. He added that he saw 'no way out'. Alternate commissioner, Akira Nakamae, then spoke. He said that Japan had seen 'this as a last opportunity' but that the IWC remained divided. He said that the way the 'discussion transpired clearly revealed dysfunctional nature and double standard at play'. Nakamae went on to say that Japan wished 'to state that there's a real risk that we'll review at a fundamental level our engagement with IWC' (IWC 59, 2007).

One thing that did occur to me, fairly strongly, is that there is a paradox in respect of resolutions. As these are passed by simple majority, they do not apparently have great weight – some resolutions are rammed through, with their promoters insisting upon votes taking place, because they know that they have the numbers necessary to win them; while their opponents, realising that they cannot prevent them, either choose not to participate in the vote or make it clear that they are not taking them seriously. On the other hand, resolutions *do* have at least a certain amount of respectability – there was no move made in 2007 to undo the St Kitts Declaration of the previous year, despite its unpopularity amongst the anti-whaling states, and the fact that it gives the pro-whaling bloc the right to argue that a majority, at least once, agreed with its views on sustainable use.

Another interesting thing, I think, is the role of the smaller states – and especially of the African states. With the exception of South Africa and Kenya (which no longer attends), they all by and large follow the pro-whaling line – even Mali, which within a few days of IWC 59 would be co-sponsoring, with Kenya, a proposal at CITES to place a 20-year moratorium on the ivory trade. This even makes me wonder whether Mali has not been 'induced' by one side (Japan) at the IWC and by the other (in particular by Western NGOs) at CITES.

Also, on at least one occasion an African country apparently voted in error – Senegal, as the first of the pro-whaling bloc countries to vote on a resolution on Non-Lethal Use of Cetaceans, voting 'yes'. Every other member of the pro-whaling bloc announced that it was 'not participating in the vote' and it seemed to me that Senegal had not understood the vote. The voting *was* often difficult to understand – on one occasion, as a vote was being taken, a delegate from a country in the Like-Minded Group asked the South African delegation which way he/she should vote. One thing that was certainly clear was that the pro-whaling bloc had approached the voting with the intention of 'not participating' in many of the votes.

I was advised that this was the first year in which this tactic of 'not participating' in votes was adopted; in my opinion, it was part and parcel of the pro-whaling bloc's approach to the meeting generally. This approach appears, to me, to have been to walk away from the meeting having lost every single point, and thereby to be able to claim the moral victory of having been stymied at every turn by an uncompromising anti-whaling bloc – and this is certainly what happened. The anti-whaling countries essentially gave not one single inch during the meeting – thereby playing into Japan's hands. Some countries, like the UK, came across as being particularly hardline. Many countries praised the US Chair from the floor for his handling of matters, the obtaining of consensus on occasion and the general good spirit of the meeting. I considered his handling to be weak, however, and thought that it was entirely due to Japan's 'manipulation' that the meeting ran with a 'good spirit' – at least for the first two days. Even in the first two days, however, there were signs that this spirit of consensus was not going to last – the sniping from Iceland being particularly telling in this regard. Norway was fairly quiet, generally – much more subdued than many others in the pro-whaling group. This certainly contributes to the impression I gained interviewing involved people in Norway – that the state has effectively given up on the IWC.[34]

In the end, much of IWC 59 was a curious mix of nervousness and braggadocio – petty point-scoring, grandstanding and playing to the gallery. Many country representatives seemed to feel that they needed to make themselves visible, and many made points from the floor, which were probably aimed more at pleasing their own citizenry than as substantial contributions to the debate. On the other hand,

34 Personal communications with Bjorn Hugo Bendiksen, Turid Rodrigues Eusebio, Rune Frøvik; Halvard Johannsen and Lars Walløe, 2007.

countries in the Like-Minded Group appeared genuinely concerned, early on, that the US bowhead quota might not be renewed – and their relief appeared genuine when this did not happen. After this, they became quite buoyed up – and some even arrogant. The countries in the pro-whaling group seemed far more orchestrated in their approach – better prepared. Many of them – with the exception of Norway, which supported but did not really enter the fray – took the opportunity to make dramatic points that would not have been out of place at a conference on racism, human rights or colonialism.

3.9 Possibilities for change

The moratorium has not prevented Japan and Norway from whaling – although neither does so on the scale on which they appear to desire. Norway objected formally in 1982 to the IWC's indefinite imposition of a nil quota and is therefore entitled to hunt commercially and does so. Norway maintains this objection. Japan did not object as Norway did, but takes whales commercially under the category of 'scientific research', which is allowable in terms of IWC rules. In fact, all of Iceland, Norway and Japan have at one time or another caught whales for so-called 'scientific research'. Member nations can set their own quotas for these programmes without consulting the IWC. Member states then issue permits to their own nationals to meet the quotas.

As well as moratoria, sanctuary areas have been promoted by the anti-whaling states as a way to protect whales. Pro-whaling IWC members have used the objection procedure here too. Japan, for instance, officially objected to the creation of a sanctuary for whales in the Southern Ocean in 1994, and is thus not obliged to recognise it.[35]

Many treaties allow reservations which, Lyster argues, are useful because they encourage states to join, and to continue to participate in, a treaty in circumstances where they are prepared to accept most but not all of its provisions. However, reservations may also provide loopholes which enable states to defend 'vested interests which conflict with the spirit of the treaty' (Lyster, 1985: 9–10). The use of the pejorative description 'undermine' is of course a value judgment. A contrast can be drawn between the ICRW and the EU's Common Fisheries Policy (CFP), which allows for no 'opt-out'. The CFP, which evolved in the 1970s as a way to ensure equal access of EU member states to each other's fishing grounds, does not permit a member state to opt out of a decision to conserve fisheries in EU waters (Rose and Paleokrassis, 1996). This does show that an international environmental treaty can be handled differently to the approach taken by the drafters of the ICRW. However, the European Union example is a strange one – involving as it does a radical giving up of sovereignty generally.

35 First proposed by France in 1992, this sanctuary was accepted by the IWC at its meeting in Mexico in May 1994. Only Japan voted against the proposal. Norway did not cast a vote.

Asked whether she felt, from the perspective of the IWC Secretariat, that there were inherent problems with the IWC, Grandy said that she thinks that the Convention is 'flexible enough to allow countries to do anything they want to do'; but that the 'problem' is that amongst the members there are 'many different views about management of whaling'. When I pointed out that Birnie has also described the ICRW as being 'flexible', Grandy said it was one of the first Conventions 'which did have a conservation aspect to it', but that the problem is that member countries do have very different views of what they want to see as priority issues (Grandy, pers. comm., 2007).

The very strong impression which I gained from interviews conducted in Norway in 2007 is that the Norwegians feel that they offered a number of significant concessions[36] in the early 1990s, that these were not accepted and that Norway is not planning again to make such an offer. Norway is, in fact, probably glad that these concessions were not accepted. At present, the country's representatives in the whaling field believe strongly that their stance is the correct one, that they are generally a country with a carefully thought-out attitude to environmental protection, and that they have an interest in not being dictated to by countries with different interests and poorer environmental records (Bendiksen, Eusebio, Frøvik, Johannsen, Walløe, pers. comm's, 2007).

This provides a good example, I think, of how difficult rapprochement is going to be to find. The landscape of the ICRW is littered with failed efforts to broker settlement and to find alternatives. Efforts remain ongoing – although somewhat sporadic. To me it seems important, however, that such efforts be continued, no matter how disheartening their failures might be for the parties who make them. Change and compromise remain possible, but for these to happen quickly is probably not. Parties need to act with patience and to build the foundations for eventual accord slowly. The compromise brokered on elephants at CITES CoP 14 in 2007 shows how compromise is possible. The conflict within the IWC runs deeper than within CITES, and it will take greater effort from the antagonists, but there are some signs of a possible cooling of the conflict.

In 2007, I asked the then South African Minister of Environmental Affairs and Tourism, Marthinus van Schalkwyk, for his impressions of the IWC. van Schalkwyk described the ICRW as being a 'very old convention' with some of its procedures being extremely outdated and strange compared to some others. Pressed for examples of such procedures, he said:

> on the floor, you know, it's actually an extremely open agenda and process where countries can come and present, you know, local communities to put their case to an international body. The volume, the load of work, of these kinds of bodies will simply become too much if that is allowed to happen. They need a much more modern approach here – and I think there was a

36 Such as conceding to the presence of international observers, paid by Norway, on whaling vessels.

> general feeling by countries who did not want to say so publicly – but who said it privately – that the body is really in need of modernisation. I was struck by the many bilaterals that we had with pro- and anti-whaling countries where people privately said we need a compromise where we must accept some of the positions of the other side in the process of compromise – but while we were on the floor there was a total unwillingness to do it. So somehow we need to break this stalemate here. I think it's possible, but it will require much stronger leadership.
>
> (van Schalkwyk, pers. comm., 2007)

In March 2008, an intersessional meeting of the IWC was held in the UK – in an effort to chart a course forward for the IWC, and to see if it might be possible to broker a compromise (IWC, 2008). According to the IWC's announcement of the meeting, at IWC 59 it was agreed by the Commission that:

> an intersessional meeting should be held prior to the 2008 Annual Meeting to provide an opportunity for Contracting Governments to discuss the future of the organisation, given inter alia the impasse reached on the RMS and the number of issues for which polarisation rather than consensus appeared to be the norm within the Commission.

At IWC 59, it had been agreed that a Steering Group, comprising the Commission Chair, the Vice-Chair and the commissioners for Chile, New Zealand and Palau, would prepare a draft Agenda. Meeting in October 2007, the Steering Group 'agreed that there is an urgent need to explore ways that might be more successful and which can improve levels of trust amongst members and others' and agreed that 'to this end, rather than launching into negotiations on substantive issues where major differences among IWC members exist', it would 'initially' be 'more fruitful to take a process-oriented approach and to seek ways to improve how negotiations within the IWC are conducted'. The intersessional meeting, the Steering Group concluded, will 'therefore focus on matters of process rather than on matters of substance and Contracting Governments will be encouraged to share their ideas on how to improve the working of the Commission'. The intersessional group was held in the UK in March 2008, and professional mediation was attempted with an outside facilitator (IWC, 2008).

As to actual results of the effort to find 'positive ways forward' for the IWC, the Chair of the IWC (William Hogarth of the US) was quoted as saying that a number of issues had been highlighted, including making better efforts to reach decisions by consensus; reducing the use of voting; adopting measures to ensure adequate notice of matters to be considered, so as to reduce surprises; employing 'cooling off' periods when difficulties arise; considering using small negotiating groups; improving coordination between the IWC and other relevant international conventions; discussing how to assist the Chair in the running of meetings; discussing how elements of civil society might be better integrated into the Commission's work; and considering whether to change the timing of Scientific

Committee meetings in order to provide more time for consideration of its reports, and to undertake review of its composition and function (IWC, 2008).

The Chair was then quoted as committing himself to producing a report of the meeting to the Commission and, in consultation with others, developing a series of recommendations to improve procedures which will 'take the Commission forward'. IWC 60 met in Chile in mid-2008. The meeting appears to have been relatively quiet, and to have reflected largely – as in 2007 – Japanese and American efforts to move toward some form of compromise. As a reflection of this, Japan stated that it would not request a vote on its proposal to relieve the 'hardship suffered by' its four community-based coastal whaling communities – Japan stating that it would not ask for a vote 'because of the progress it saw in the discussions related to the future of the IWC' (IWC 60). As other outcomes of the meeting, the Commission agreed that it would take steps to 'improve the practice and procedures of the Commission' and to 'further discussions/negotiations on substantive issues'. Also, a proposal on the creation of a South Atlantic Sanctuary, put forward by Argentina, Brazil and South Africa, was withdrawn – the three countries indicating that they would not ask for a vote on the issue 'in order to facilitate the work on the future of the IWC' (IWC 60). Arising from the intersessional meeting and IWC 60 in 2008, approaches were agreed by consensus to reforming the IWC's working procedures and practices, and it was agreed that there would be further discussions and negotiations on substantive issues. In respect of working practices, it was agreed that every effort would be made to reach consensus on all matters of substance, and that voting would remain a last resort.

After IWC 60, a small working group of 25 countries, including all of the major players, met in the US in September 2008 to consider a 'package deal' that might go so far as to include a limited resumption of whaling, and the bringing of small cetaceans under IWC control. A second small working group met in the UK in December 2008 and then in Italy in March 2009. Between 2007 and 2010, then, the parties engaged in a process which was eventually given the official title the 'Future of the IWC', in which compromise was sought through the acceptance of a 'package deal' which would include compromise on issues such as Aboriginal Subsistence Whaling (ASW); Japanese Small Type Coastal Whaling (JSTCW); the lifting of the 1982 'moratorium' on commercial whaling; the management of small cetaceans; and the creation of whale sanctuaries, amongst others. IWC 61, 2009, was, therefore, a very quiet meeting and it was agreed by consensus that the process would be extended to IWC 62 in 2010. A Support Group[37] was formed to assist the Chair (now Cristian Maquieira, from Chile), and the 25 countries met in March 2010 in the US, after which the Chair released a document called 'A Proposed Consensus Decision to Improve the Conservation of Whales'. It was then envisaged that the 'package deal' would be considered for adoption at IWC 62 in 2010.

37 Comprising Antigua and Barbuda, Australia, Brazil, Cameroon, Germany, Iceland, Japan, Mexico, New Zealand, St Kitts and Nevis, Sweden and the US.

The Proposed Consensus Decision began with a 'vision statement' to the effect that the IWC would continue to 'work cooperatively to improve the conservation and management of whale populations and stocks on a scientific basis and through agreed policy measures', and that 'by improving our knowledge of whales, their environment, and the multiple threats that can affect their welfare' the IWC 'will strive to ensure that whale populations are healthy and resilient components of the marine environment'. The Decision then recorded that 'the status quo is not an option for an effective multilateral organisation' and that 'to overcome the present impasse, the IWC has in recent years recognised the need to create a non-confrontational environment within which issues of fundamental difference amongst members can be discussed with a view to their resolution' (IWC 2010).

The fundamental components of the Decision were that the IWC would retain the moratorium on commercial whaling; suspend for a ten-year period unilaterally determined whaling under special permit, objections and reservations; bring all whaling authorised by member governments under the IWC's control; limit whaling to those members currently taking whales; ensure that no new non-indigenous whaling takes place on whale species or populations not currently hunted; establish caps for the next ten years that are significantly less than current catch levels and which are, within sustainable levels, determined using the best available scientific advice; introduce modern and effective monitoring, control and surveillance measures by the IWC for non-indigenous whaling operations; create a South Atlantic Sanctuary; recognise the non-lethal value and uses of whales, such as whale-watching, as a management option for coastal states and address related conservation, management and scientific issues; provide a mechanism for enterprise and capacity building for developing countries; focus on the recovery of depleted whale stocks and take action on key conservation issues, including by-catch, climate change and other environmental threats; set a decisive direction to the future work of the IWC, including measures to reform the governance thereof; and establish a timetable and mechanism for addressing the fundamental differences of view amongst members in order to provide for the effective long-term functioning of the IWC (IWC 2010).

At the meeting, however, consensus on the Decision proved impossible to find, despite intense discussions, and the Chair eventually ruled that the IWC had not found consensus. The Chair then, probably in a desperate effort to save the process, suggested that a 'further period of reflection' was required, and this was agreed to (IWC 62, Chair's Report). This meant that although no formal intersessional steps were taken, the parties arrived at IWC 63 in Jersey still discussing the matter. On the other hand, the general feeling in the air appeared to me to be that it would not be easy to revitalise the process. Dan Goodman told me in mid-2011, at IWC 63, that Japan was disappointed as after spending several years trying to reach compromise, the package had failed because no compromises had been offered by the anti-whaling side. His words were that the pro-whaling side had 'offered huge compromises' but that 'nothing came from the other side' (pers. comm., 2011).

3.10 Things almost fall apart

IWC records after IWC 63 state that the parties agreed to maintain the progress achieved through the 'Future' process and that there was general agreement to encourage continuing dialogue regarding the future of the IWC; to continue building trust by encouraging the coordination of proposals or initiatives as widely as possible before submission; and encouraging continued cooperation in taking forward the work of the IWC, notwithstanding different views regarding conservation of whales and management of whaling (IWC 2011). However, merely reading of this agreement to continue the 'Future' process disguises that IWC 63 was an alarmingly acrimonious meeting – to the extent that the entire Commission seemed at several stages on the point of complete collapse.

The meeting began with a crisis of leadership as the Chair had resigned during the previous year before expiry of his term, as had the Vice-Chair. During the preliminary week of committee meetings no commissioner expressed willingness; but eventually, under strong pressure from all sides, Herman Oosthuizen from South Africa agreed to be interim Chair of what proved to be a very difficult meeting. IWC 63 provided fascinating insight into how issues of substance and of procedure can overlap. In my view, the meeting involved a bitter 'battle' for control of the IWC, with this battle being fought out not overtly on substantive issues, but through issues of procedure. Four issues dominated debate, these being the non-attendance of certain delegates due to their failures to obtain visas timeously from the UK to visit the States of Jersey; whether a proposal properly submitted by a contracting government could at the meeting itself be submitted by a different party; whether to allow contracting governments to pay their annual subscriptions by any means other than electronic bank transfer from a bank account held in the name of the contracting government; and, finally, how a voting quorum is constituted in the IWC.

The first of the four issues saw a number of contracting governments affected, including Russia and Senegal. There was no evidence that this was a deliberate ploy by the UK government to exclude certain delegations, despite this being at least insinuated in plenary by some commissioners. Ultimately, the issue was resolved through an investigation and report by the Secretariat; general discussion; an undertaking from the UK to work with the Secretariat to ensure that the situation did not again occur; and agreement by the Commission that the IWC Secretariat and the host country of annual meetings should take a number of steps, including provision of detailed information to members well in advance of the meeting to help visa acquisition. Although, obviously, the issue was potentially a serious one, it was arguably used more as a delaying/distracting tactic. If a vote had been held and delegates from contracting governments who would have wished to be there had been unable to be present because of a visa delay, the issue would have been serious. In the end, however, it petered out.

On the second issue, according to the IWC's Rules of Procedure,

> no item of business which involves amendment of the Schedule to the Convention, or recommendations under Article VI of the Convention, or

> Resolutions of the Commission, shall be the subject of decisive action by the Commission unless the full draft text has been circulated to the commissioners at least 60 days in advance of the meeting at which the matter is to be discussed.

A proposal had been so put forward ('circulated') by the UK more than 60 days before IWC 63. However, at the meeting itself, the UK indicated that it would prefer not to put the proposal forward under its name. Instead, the proposal was to be put forward in the name of the European Union. Various other contracting governments objected, however; in particular Russia (which asked why the European Union should have any more status than the World Boxing Association in the IWC) and St Kitts and Nevis. Objections included that a party which had not put forward the proposal originally could not at the meeting itself become the proposer, and that the European Union was not a contracting government to the ICRW. After much debate, a compromise was reached and the proposed resolution, this being a resolution on 'Improving the Effectiveness of Operations within the IWC', was recorded as having been put forward under the heading 'submitted by the United Kingdom and Poland, Belgium, Czech Republic, Denmark, Germany, Estonia, Spain, France, Italy, Luxembourg, Hungary, the Netherlands, Austria, Portugal, Finland and Sweden'.

The clause requiring that contracting governments pay their compulsory annual subscriptions by EBT, from a bank account held in the name of the contracting government, proved to be the most contentious of all of the clauses in the resolution, despite having being repeatedly described by the UK as a 'modest proposal'. The UK expressed its surprise that the matter was even controversial, and argued that all it was trying to do was to modernise the IWC, and obviate the need for the Secretariat to handle large sums of money in different countries. At the same time as the UK delegation inside the plenary was arguing this, however, media reports outside suggested that the object of the proposed resolution was nothing less than significant reform of the IWC – in particular, by preventing corrupt practices. It was written, for instance, that 'reforms to prevent "votes for cash" allegations were set to top the agenda' and that 'the most eye-catching of the UK reform proposals is that governments should have to pay their membership subscriptions by bank transfer, creating an auditable trail' (Black, 2011). The UK's Environment Minister, Richard Benyon, was quoted as saying that the IWC 'has been going since 1946, and it needs to modernise its procedures so it doesn't leave itself open to the kinds of allegations made a year ago' (ibid.). The UK's protestations of 'surprise' were somewhat undermined by such pronouncements, by the importance which it attached to the clause and by the obduracy with which it insisted on the proposal being voted upon. Ultimately, the UK knew that a majority of the parties present would support it, and that only a simple majority was required.

In the end, however, a compromise was reached after a small working group had negotiated. The final text read as the UK had proposed it and insisted it should be. The compromise came in in the adoption of a different, but related, clause which provided that for the purpose of application of the rule, payments of membership dues will count only when the funds have been credited to the Commission's account 'unless the payment has been made and the Commission is satisfied that the delay in receipt is due to circumstances beyond the control of the Contracting Government' (IWC 63, Resolution 2011–1).

That left the final issue, and it proved the most controversial of all. Although the IWC has for many years been a byword for conflict, 2006 was the last year in which there was open conflict – and 2007 was the last year in which any issue was voted upon. For approximately a decade (2001–2008), a regular proposal was put forward annually, by Argentina, Brazil and South Africa (range states), for the establishment of a whale sanctuary in the South Atlantic. In most of the years the proposal was put forward it was successfully voted for by a simple majority – the last year of voting being in 2007. The proposal was, however, withdrawn in 2008, and was not put forward in the years 2009 and 2010, as it was part of the proposed 'package deal'. In 2011, however, Argentina and Brazil put a similar proposal forward once again. Putting the proposal forward was controversial as it required separating the proposal from the 'package deal' which was part of the 'Future of the IWC' process.

Japan noted that in previous years (until 2007) it had requested a vote on its proposal for an interim relief catch allocation to relieve hardship; but advised that 'in a spirit of co-operation' it had not requested a vote during the discussions on the 'Future of the IWC' process. Japan advised that, although disappointed at the outcome of those discussions in 2010, it 'recognised the improved atmosphere of discussions that process had engendered'. Finally, Japan advised that, 'appreciative of the efforts for consensus that had already occurred in discussions, it again would refrain from asking for a vote this year'. Various parties expressed great reluctance to have the South Atlantic sanctuary proposal voted upon, and requested that it be withdrawn. Argentina and Brazil, supported by other members of the 'Buenos Aires Group' (Chile, Colombia, Costa Rica, Ecuador, Mexico, Nicaragua, Panama, Peru and Uruguay) insisted, however, upon their right to put a proposal forward and to insist upon a vote – several EU countries gave strong support. Various parties, on the other hand, characterised the insistence on putting the proposal to a vote as leading potentially to the 'destruction' of the IWC's 'Future' process – Russia, in particular, making this point strongly.[38] One argument made in this regard was that putting the proposal to a vote might lead to other issues being put back on the agenda singly in future years (such as Japan's small-type coastal whaling) and to issues being linked politically (such as JSTCW and aboriginal subsistence whaling).

The 'sustainable use' group parties did not initially join the meeting, and, when they did, made no secret of their intention to leave the room en masse and break the quorum if a vote was insisted upon. After debate, Argentina and Brazil

38 According to the Chair's Report, in plenary debate Russia requested withdrawal of the proposal 'so as not to destroy the process of the Future of the IWC' (IWC 59, Chair's Report). According to my notes, the Russian Commissioner said (in translation) 'this proposal kills the Future of the IWC by inhumane methods'.

continued to insist upon a vote. The Chair then instructed the Secretary to prepare for a vote. The 'sustainable use' group parties then exited the room, leaving less than half of the Commission's contracting governments represented. Various Buenos Aires group and EU countries insisted, however, that the meeting remained quorate on the basis that it had been quorate at the beginning of the meeting.[39] The consequences of the meeting being quorate were important – were the meeting still quorate and a majority present to vote in favour of the sanctuary being created, then the argument could be made that the sanctuary would have come into existence. Of course, that would then have led to years of further debate on the procedural legalities, objections and reservations being lodged, strong accusations of bad faith and illegality being made and possible withdrawals from the IWC.

The Chair then adjourned to a series of Private Commissioners' Meetings where debate continued on the meaning of 'quorate'. According to the IWC's Rules of Procedure, 'attendance by a majority of the members of the Commission shall constitute a quorum'. The rule is stark – unfortunately, it is this very simplicity which leaves it open to different interpretations. After nine hours of Private Commissioners' Meetings on the final day of the Meeting,[40] including numerous meetings of sub-groups, a compromise was reached. No resolution was taken, but certain wording was agreed to which would be included in the Chair's Report.[41]

To the casual onlooker, it might seem as though IWC 63 became bogged down in procedural matters and that the various parties involved lost sight of the main goals of the Convention. This would be, it is submitted, too simplistic an explanation. While the bulk of the meeting did involve procedural matters or discussions as to form, and actual whale conservation or management issues were little discussed, the procedural issues were not unimportant. Strongly arguably, what was at stake was control of the IWC. IWC 63 was important in this regard, but not the final word.

39 The UK made this argument strongly, and was even to continue to make it in written submissions to the IWC over the ensuing months – despite the logical problem that if a meeting remains quorate throughout after having been quorate at the beginning, the meeting would be quorate if only one party was in the room at the time of a vote.

40 The debate in the Private Commissioners' Meetings was not recorded and although I attended I will not comment on what was discussed.

41 The wording agreed to reflected that Japan, Cambodia, Cameroon, Côte d'Ivoire, the Gambia, Iceland, Norway, Nauru, Mongolia, Mauritania, Guinea-Bissau, Grenada, Kiribati, Morocco, Korea, Ghana, Palau (despite this record, Palau did not in fact leave), Togo, Tuvalu, St Kitts and Nevis and St Lucia had left the meeting room after the Chair instructed that a vote on the proposed South Atlantic sanctuary should begin. It was then reflected that after extensive informal consultations, there was no agreement on how to interpret the quorum rule, but that it was decided to establish an intersessional group to consider the issue. The Commission then resolved to discuss the sanctuary as the first substantive agenda item at IWC 64, and resolved that, if consensus could not be reached on the item, a decision would be taken in 'accordance with the Commission's Rules of Procedure'.

3.11 Things go back to 'normal'

IWC 64, 2012, was held in Panama. During the year leading up to the meeting, a process had been run in terms of which contracting governments were invited to make submissions as to their understandings of the quorum requirements and a small number did so; also, comparative practices from other conventions were collected. During the preliminary week there was much discussion over how this procedural impasse might be resolved, but in the final analysis the issue simply fell away. Several votes were held, all were clearly quorate and no party tried to break quorum. The first vote was on the proposed creation of the South Atlantic Sanctuary, proposed by Brazil and co-sponsored by Argentina, South Africa and Uruguay. The proposal received 38 votes for and 21 against, with two abstentions, and the requisite 75 per cent was therefore not obtained. The vote was held quietly, with none of the acrimony of the previous year. Generally, the atmosphere at IWC 64 was far calmer than at IWC 63. The commissioner for Switzerland, Bruno Mainini, agreed to serve as interim Chair for the meeting. (At the end of the meeting, Jeannine Compton-Antoine of St Lucia was elected as Chair going forward, with Frederic Chemay of Belgium as Vice-Chair.)

Although the atmosphere was calmer, I think that the meeting did become very difficult. In a real sense, the 2011 meeting ended with the anti-whaling parties firmly in the ascendancy and in control. They took a firm grip on the administration of the IWC and entrenched their numeric control, as shown in the UK's refusal to shift on having their proposed administrative reforms voted on. The groupings of EU and Latin American parties appeared to be both unified and intransigent. The 'sustainable use' group, on the other hand, came out of the 2012 meeting stronger than before – due both to a hardening of attitudes and of alliances amongst pro-whaling parties, and to cracks which appeared in the anti-whaling alliance.

The main fissure which emerged concerned the issue of ASW. Five years after IWC 59, the Russian and US bowhead and gray whale quotas were up for renewal. These were not particularly contentious in and of themselves and would have sailed through comfortably by consensus, had the parties not linked their quota applications to that of St Vincent and the Grenadines. A number of parties, particularly from Latin America, had difficulties with approving the request from St Vincent, raising concern over whether the hunt ought to be considered aboriginal. On the understanding that by the end of the meeting the decision would have been taken to meet biennially instead of annually, the period of the quota[42] became six years rather than five. As there was no consensus, and Russia and the US did not agree to separate the proposals, a

42 For the US, no more than 336 bowhead whales for the period 2013–2018; for Russia, not more than 744 gray whales, 2013–2018; and for St Vincent not more than 24 humpback whales, 2013–2018.

vote was held – in which the joint proposal was approved by 48 in favour, ten against,[43] two abstentions[44] and one not participating.[45]

Obviously, it was a 'split in the ranks' that the Latin American block of parties voted against the Russian and US ASW quota proposals, although understandable in the circumstances. The bigger split was to come from the Greenland ASW proposal. Denmark, for Greenland, sought an increased quota, over even that which had been proposed at the start of the meeting, of bowhead, fin, humpback and minke whales. The matter was postponed several times, and ran as an undercurrent through the meeting. In plenary debate it was obvious that there was much division – especially as some parties argued that by seeking an increased quota Denmark/Greenland was in breach of an undertaking made at IWC 62 in 2010. In response, Denmark relied on the argument that the increased quota was needed. Denmark eventually insisted on a vote on the amended proposal, without an alternate vote on the original proposal – meaning that if the vote was lost, the ASW quota for Greenland would be lost. The vote ran 25 in favour, 34 against, with three abstentions.[46] In responses to the vote, various 'sustainable use' group parties lamented the lost quota – Japan, for instance, saying that it was a 'sad conclusion which deprived people living in harsh climates of an important source of nourishment'; St Lucia 'noting' that as the previous quota had expired 'what the IWC had actually said to Denmark/Greenland was to go forth and manage their fishery on their own'; and St Vincent saying that the IWC 'had failed to understand the difficulties faced by coastal peoples' and added that the IWC was now 'about whales, not people'. Denmark then said that 'a number of countries had not wished to take responsibility for whaling in Greenland'; that it 'regretted the lack of solutions and the lack of understanding'; that 'this was a serious situation given the IWC had to work hard to survive at a time when it had not been able to fulfil its obligations under the Convention'; and that Denmark would now 'return home to make a sensible decision as to its future course of action' (IWC 64, Chair's Report).

Clearly this issue was provoked by Denmark choosing to go 'all or nothing' on its increased proposal, and the vote paved the way for either a formal withdrawal from the IWC or a deliberate flouting of the IWC by Greenland's whaling outside of the auspices of the IWC. At the time of writing, it was unknown which way this might go, but the least that can be said is that the unity in the anti-whaling ranks seen in 2011 is now at risk (if not gone). Another possible rift within the

43 These being Argentina, Brazil, Chile, Colombia, Costa Rica, Dominican Republic, Ecuador, Gabon, Peru and Uruguay.

44 India and Monaco.

45 The Czech Republic was absent from the room.

46 The countries which voted in favour were Antigua and Barbuda, Benin, Cambodia, China, Denmark, Ghana, Grenada, Iceland, Japan, Kiribati, Korea, Laos, Mongolia, Morocco, Nauru, Norway, Palau, Russia, St Kitts and Nevis, St Lucia, St Vincent, Tanzania, Togo, Tuvalu and the US. Voting against were Argentina, Australia, Austria, Belgium, Brazil, Chile, Colombia, Costa Rica, Cyprus, the Czech Republic, Dominican Republic, Ecuador, Estonia, Finland, France, Gabon, Germany, India, Ireland, Israel, Italy, Luxembourg, Mexico, Monaco, the Netherlands, New Zealand, Panama, Peru, Poland, Slovenia, Spain, Sweden, the UK and Uruguay. Oman, South Africa and Switzerland abstained.

anti-whaling camp arose in respect of a draft resolution put forward by Monaco on the issue of highly migratory cetaceans on the high seas. The resolution essentially directed the IWC to take the debate on these species (both those managed and those not managed by the IWC) outside of the IWC and into the realm of the wider United Nations system by 'setting up effective synergies and coordination between the IWC and relevant UN processes' (IWC 64, Draft Resolution, Monaco). Various 'sustainable use' group parties, such as Japan, spoke in opposition to the proposal, and various anti-whaling parties gave cautious support. Monaco eventually withdrew the proposal; probably, my guess is, because it had not received sufficient support from other anti-whaling countries, both in and outside of the Efuropean Union.

Japan put forward a proposal to amend the Schedule to allow for a small-type coastal whaling quota of minke whales, this being a regular proposal but one not made for some years during the 'Future of the IWC' process. Japan explained the needs of its coastal peoples, explained the proposal and listened to debate on the matter, before withdrawing the proposal – explaining that it would put forward a new proposal at the next IWC meeting, in hopes of resolving the matter.

Scientific permit whaling was discussed, with various parties presenting arguments for and against the value thereof – as expected. There was, however, a surprise when Korea announced that it was considering introducing scientific permit whaling in its waters in order to research different whale stocks which (it claimed) is difficult to do through non-lethal sighting surveys – and in response to ongoing complaints by Korean fisherman. Korea explained that it would put forward a detailed research plan to the next IWC Scientific Committee meeting. Whaling under scientific permit is, of course, extremely contentious and is the subject of Australia's current ICJ litigation against Japan. For Korea to introduce the practice at such a time must be seen as a deliberate slap across the face of the anti-whaling parties, and to be part of a general hardening of attitude by the 'sustainable use' group. Korea has, however, been a relatively neutral party for many years, and in January 2013 advised formally that it would undertake research using non-lethal methods.

3.12 Considering the position

According to Andresen, the main pro-whaling countries (Iceland, Japan and Norway) have adopted different strategies. His assessment is that Japan has been:

> most *loyal* in the sense that it withdrew the objection to the moratorium, conducted only scientific whaling, and remained in the IWC. Iceland chose the *exit* strategy, due to frustration with developments in the IWC. Norway has loyally remained in the IWC but made its voice heard by resuming commercial whaling.

Andresen then suggests that Iceland and Japan seem to be in much the same position, not being able to 'hammer out a consistent high-level strategy to deal with the issue'. Compared, he continues, 'to the much larger and still very unified anti-whaling coalition, the pro-whaling coalition has been loose and fragmented' and then argues that anti-whaling forces have been successful in that both Iceland and Japan have been split domestically, and their 'fear of negative economic consequences is greater than the wish to resume whaling' (Andresen, 2001: 245, 246–7, 247). Along the same lines, Stone describes Friedheim as maintaining 'that whaling proponents have failed to make standard bargaining offers and threats' and having described the Japanese as having failed – for whatever reason – to orchestrate normal coalitions with their natural allies, including Iceland and Norway (Stone, 2001: 277–8).

As with everything else concerning the IWC, there seems little consensus even on the negotiating skills of different parties. Victor comments that Iceland has been less able after the Cold War ended to link issues such as this to other matters with which the US is concerned (like NATO bases), whereas Norway's more 'prominent' role in world politics has 'better insulated it from US pressure'. Japan, he says, has 'tried hard, but failed, to act strategically through the IWC, failing to stop the adoption of the Antarctic whale sanctuary'. He points out that Japan has failed also in its efforts to have its coastal communities, which have long depended on small-type shore-based whaling, given the same rights as aboriginal communities, which would have allowed them to continue limited whaling for local consumption. Friedheim, according to Victor, has 'convincingly documented that many of these Japanese failures were the result of poor strategies and tactics by Japanese negotiators within the IWC, exacerbated by divided positions of the Japanese government and perhaps anti-Japanese and racist attitudes within the IWC'. His conclusion on this point is that Japanese delegations have pursued a strategy based on science and aimed at adopting sustainable management quotas, but argues the IWC has been, since 1982, a highly politicised body for which science has been increasingly less relevant (Victor, 2001: 300).

The charge of Japanese naïvete seems implausible to me – the Japanese would, rather, appear to be the master strategists of environmental debate. They have, after all, been 'whaling and dealing' within the IWC since 1951, as well as fighting environmental battles in many other arenas. However, when I asked Kleinschmidt for his opinion on this, he considered the Japanese to be bad, or clumsy, negotiators. Asked whether the Japanese are 'the real masters of diplomacy', he said 'no, their diplomacy's poor, through a serious failure to understand other peoples' debates' (Kleinschmidt, pers. comm., 2006). According to Day, in Japan political intrigue seems to be 'essentially viewed as an art form; and as such, in the sphere of whaling issues in the Third World, it seems to be Japan's primary cultural export item' (Day, 1992: 127).[47] This would appear to be the more plausible view of

47 Day writes that in 1981:

> Brazil abstained from the moratorium vote, and informed the IWC that it was getting out of the whaling business that year. In 1982 it not only remained in the business as a Japanese whaling colony but changed its abstention to a no vote against the moratorium. The Brazilian newspaper *Folha de S_o Paulo* made it absolutely clear why the decision had been made. Japan had offered Brazil a $400-million agricultural investment programme.
>
> (1992: 132–4)

Japan, although as usual the truth probably lies somewhere between the alternatives. Further, in the light of such comments as those of Day's, is it far-fetched to see the Japanese as linking the ivory trade to the resumption of commercial whaling and to see them deliberately playing CITES and the IWC off against each other?

In recent times the anti-whaling camp has been in the ascendancy, although not with an overwhelming majority until very recent years, in which its numerical superiority has been clear. Friedheim concedes that 'the existence of a seemingly powerful majority and a weaker outvoted minority might imply that the majority has won all of its objectives, the minority lost all of its rights'; but then contends that that is not 'quite true'. He argues that:

> the majority has won the big votes, created a moratorium they refuse to lift, and augmented it with a Southern Ocean sanctuary, but they have not been able to force or convince the minority to give up all of their rights to whale. In short, preservation as a new norm has not been internalized by key states to whom it is intended to apply.
>
> (Friedheim, 2001b: 10–1)

According to some, it is essential that the disputes within the IWC be resolved. Gambell, for instance, suggests that the 'contentious and potentially divisive issue' of whether the ICRW is to be used 'as an instrument to implement a new environmental conservation ethic' must soon 'be resolved lest the IWC forfeit its position as the global authority for management of whales and whaling' (2003: 85). It has also been argued, on the other hand, that in fact the present situation is already as close to being ideal as it is likely ever to be. Victor, for instance, writes that promoting international law and governance could be 'a compelling reason to reform' the IWC, but that 'building a new agreement will result in either deadlock (and thus no agreement) or extensive reliance on US power (and thus a coerced agreement)' – he asserts that neither would be good for the rule of law. However, he feels that the 'present messy outcome' might even be what is known as 'Pareto optimal' – a situation where 'no party could do better without harming the others' (Victor, 2001: 303–4, 304).

Questioned on this, Oosthuizen felt that it was 'the crux of the state of the IWC at the moment', with some countries apparently being content with the current situation; but that on the other hand perceptions could be deceiving with some countries seeming to be actively involved in the management and/or conservation of whales, while at the same time most likely undermining the proper functioning of the IWC. He pointed out that more whales are being killed now under scientific whaling than would have been the case if there were commercial whaling, and argued that Japan is 'loading their dice' by proposing the scientific take of humpback and fin whales, and by increasing the number of minke whales taken. The question, he asked, is why: with one possibility being that it is to start negotiations from a strong position, and another that it might be to stretch the credibility of the IWC to the point where it 'snaps' (Oosthuizen, pers. comm., 2006).

According to Friedheim, however, neither side is willing to accept the present situation as a permanent outcome. He bases this judgement on the argument that the 'preservationists are not willing to concede what the whalers now "have", and the whalers insist upon having more than what they have been allotted under the present regime'. Both, he says, 'push toward improving their situations and force us to consider, if they continue, whether the IWC will remain a viable international resource management organization' (Friedheim, 2001a: 311–2). Friedheim then suggests that, increasingly, an:

> important part of the IWC's annual business involves passing resolutions concerning all aspects of whales and whaling in order to provide guidance for (if one wishes to be polite) or to put pressure on (not polite) the parties to conform their behaviour to the resolution's demands.
>
> (2001c: 206)

Such resolutions, he says,

> have no formal standing in international law, and in theory they are not binding upon the parties. But those who endorse such resolutions hope they will become part of 'soft law', purportedly reflecting commonly accepted norms, and thus will be obeyed.
>
> (ibid.)

Friedheim, I think, gives a strange interpretation of 'soft' law. 'Soft' law has very little status. Rather, the hope of the 'endorsers' must be that the resolutions will eventually harden into 'hard' law – and then be binding as customary law.

Criticism within the IWC can be extremely harsh. Lapointe avers that the situation within the IWC appears to be 'unchanged, unchallenged, and of no concern to many of its commissioners'. Membership, he cautions, of an international organisation 'is a serious matter, requiring adherence to the legal provisions and treaty assurances that there will be no encroachment on the sovereign rights of other members'. Even more caustically, he suggests that there are 'numerous indications that certain nations (Mexico, New Zealand, the UK, and the US, for instance) have simply forgotten the full meaning of becoming a member of an international forum', and as an example he explains that the 'honest observer cannot have the smallest doubt that the IWC's decision to establish a Southern Ocean (Antarctic) Sanctuary was made in blatant disregard of the legal provisions of the ICRW' (Lapointe, 2003: 56).

Burke is pessimistic, suggesting that the IWC 'has no direct role to play, even if the aim of interested states is to modify the ICRW itself'. This is because the ICRW 'has no provision for amendment at all, let alone giving some role to the commission', and 'hence the commission itself has no authority to change the basic treaty' and instead its parties must negotiate a new agreement. He concludes that there would be formidable philosophical difficulties standing in the way of achieving this (Burke, 2001: 51–2). Burke is incorrect on this legally – the text of certain provisions of the ICRW was amended once, by a Protocol of 1956, and could be again. On the political issue, he is probably correct.

Kleinschmidt, a former Vice-Chair, said that the IWC 'tries to modernise itself, it tries to reinvent itself'; but that he sees that merely as 'an attempt to restate its 1946 mandate' and that the IWC 'has all the appearance of a tired organisation'. He explained that by 'tired' he meant 'in the sense that it is totally ramified in the technical debate around the moratorium-or-not issue'; it is, he said, 'all again words, and the outsider wouldn't really easily understand' with 'the actual debate that needs to be had never surfacing there' (Kleinschmidt, pers. comm's, 2006, 2008).

Burke argues that it is neither defective drafting nor indeterminate terminology that has led to distortions in the implementation of the ICRW, the treaty itself being 'straightforward in expressing the purpose of conserving whales in order that species and stocks can continue to be harvested' and nothing within it being so ambiguous that it can 'reasonably be made to yield an interpretation precisely the reverse of the original purpose'. His conclusion, however, is that 'unfortunately, no drafting legerdemain can successfully prevent the regime destruction that results from the blindness of ideological conviction and moral superiority' (Burke, 2001: 59–60). The question might be asked: does the argument then become one as to whether subsequent practice can change the original purpose or not?

On the issue of whether members are generally getting what they want, Freeman contends that it would be naïve not to recognise that 'for the majority of participants in IWC discussions, the current non-resolution of the whaling problem is the desired outcome' (2001: 125). This may not be the worst thing, however. According to Vogler, Pareto optimality can be explained as a situation where a change would make some people better off but without at the same time making anyone worse off. This is not, he says, 'the same as fair or equal shares to a common resource and would often be politically unacceptable because it does not deal with the relative deprivation felt by those whose welfare is not improved' (Vogler, 2000: 19).

The argument is supported by Ishii and Okubo, who posit that Japan might not be serious about trying to resume commercial whaling – they argue that it is 'logically possible' that Japan, by claiming the opposite, may be obscuring its real diplomatic objective. They suggest that the 'primary objective' of Japanese whaling diplomacy (led by the Fisheries Agency of the Ministry of Agriculture, Fisheries and Forestry, together with the Institute for Cetacean Research and the Japanese Whaling Association) might be the continuation of scientific whaling. They put this very strongly, arguing that the whaling camp does not want the moratorium on commercial whaling to be lifted, and that it is pretending to be devoted to resuming commercial whaling because it prefers the status quo (Ishii and Okubo, 2007).

They then argue that Japan has not pursued any of the obvious strategies that might enable it to resume commercial whaling. First, Japan has done little to create a favourable atmosphere for negotiation within the IWC; in fact, they suggest, Japan appears intentionally to have created an acrimonious atmosphere. Second, Japan has done little to respect the scientific community and achieve respect for Japan's scientific credibility; instead, Japan has undertaken much research irrelevant to management advice, and has in fact done much to undermine the work of the Scientific Committee (and by denouncing as biased scientists who have opposed the Japanese line). Third, Japan has done little apparently in earnest to negotiate with the anti-whaling nations. Ishii and Okubo give several examples of this apparent disinterest, such as the retreat from their apparent rapprochement with the US over bowhead whaling quotas in 2002. The authors point out that Japan did not foster links with the US after 2002, despite the US having – at the intersessional meeting of the IWC – supported Japan's efforts to obtain interim relief allocations of minke whales for their four coastal whaling communities.[48] Fourth, the authors suggest that Japan has done nothing to create a strategy, a support network, which would enable the country to leave the IWC – despite its numerous threats to do so. Ishii and Okubo outline two possible reasons for Japan acting as they suggest it is doing. First, they propose that the explanation may be rooted in domestic politics, with the Fisheries Agency of the Ministry of Agriculture, Fisheries and Forestry being desperate not to give up the subsidies and commissions that accompany research whaling. Second, they suggest that Japan's stance on whaling might give the country domestic (and even international) credibility by showing that it can 'stand up to' the US – even though since 1945 Japan has 'very seldom been able to deny the US anything' (Ishii and Okubo, 2007: 55–87).

While I can agree with the general logic of their argument, it appears to me that they simply do not take their argument far enough. There remains at least one more possible – and not implausible – explanation for Japan's being 'content' with the status quo, which is the possibility that the entire whaling issue is simply an extremely large 'red herring' which serves to distract attention from the real damage which Japan is doing, in its short-term interest, to other areas of the global environment – fisheries in particular. In other words, while the world focuses its attention on whaling – with its ongoing feints, lunges and parries – another environmentally damaging use of natural resources is kept largely below the radar.

Goodman has suggested that 'it is likely that' the current position 'serves the domestic political agendas of its contracting governments better than any revision acceptable to all parties would do' (1997: 547). One problem, however, with the contention of Pareto optimality is simply that none of the parties currently *appear* to *be* content with the way things presently stand. Victor points out that although the present system 'serves the interests of most participants, no party achieves all that it wants', and that some 'are even seriously injured as fundamental principles of justice and human rights are violated' (2001: 299). It may also be that states are less content with the current position, than that they are scared to change anything in case they lose thereby. Friedheim argues that:

> the strategy and tactics used by both pro- and anti-whaling states, state coalitions, and nongovernmental organizations has produced stalemate, and

48 It is probably unfair to blame Japan for a 'retreat' here, as it is more likely that the US reverted to its traditional stance after having obtained what it wanted.

> unless some of the major players break from their long-held positions, stalemate will be the best that can be hoped for in the foreseeable future.
>
> (2001c: 200–1)

This is not, per Friedheim, 'to say that the major actors are happy with the present stalemate, but they fear they will be worse off if they shift their position or take another approach to the problem' (ibid.).

It has been suggested that even if one side is *not* dominating within the IWC, the perception that this is happening can itself be damaging for international law. Per Stone, it must be troubling that there is 'the unmistakable *appearance* that the IWC has been kidnapped' – whether the commission is legally ultra vires or not (2001: 273). Victor writes that principles of justice are not completely ill-served by the present system and might fare worse under a reformed body, and that stopping whaling would require instruments such as sanctions and boycotts which would trample rights further. Probably, he continues, 'it is worth reconciling environmental and trade conflicts on a more important issue for animal rights advocates and environmentalists (and the animals themselves), such as the protection of truly endangered species' (Victor, 2001: 305–6). This is an argument to be considered, but, given that whales and elephants are symbols, how could killing them with global approval persuade persons to conserve generally? Only by fully embracing the CBD, perhaps, as well as educating more people than it is possible to imagine reaching.

Victor argues further that the 'risk of missteps is greatest in the whaling states, especially Japan, which incorrectly thinks that it will benefit from the formal resumption of commercial whaling. It appears that Japan still does not know how the game should be played'. As an example of this, Victor refers to Japan's decision in 2000 to include 'charismatic Moby-Dick sperm whales' in its scientific whaling programme, and suggests that this decision 'earned scorn across the West and forced the US government to renew threats of sanctions' (Victor, 2001: 307). However, this decision – which was to take ten sperm whales – must be seen as a slap of glove across face. To describe the Japanese as 'not knowing how to play the game' is arguably quite bizarre. Probably Japan is quite aware of how far it can push the US before sanctions will be instituted.

Iino and Goodman suggest that:

> Japan's policy towards the IWC is best understood as a strategic effort to encourage the IWC to function in accordance with the objectives and provisions of the ICRW as well as relevant principles regarding conservation and management of marine living resources, such as sustainable use based on scientific evidence.

In response, the writers suggest, 'to recent scientific evidence concerning the interaction between cetaceans and fisheries, Japan is also encouraging the IWC to adopt an ecosystem approach to the management of marine resources' (Iino and

Goodman, 2003: 6). In other words, whaling is just part of the whole – Japan may be after far more than a few more whales on restaurant plates.

Concurring at least partially with this, Friedheim writes that there is:

> pressure from the flank and that the restoration of limited trade in elephant products is one reason I would not include a strict trade ban in an IWC reform. Rather, it seems to make sense that if the limited elephant product regime is successful, it should provide a set of practices that the IWC can emulate. In any case, I expect that what has been happening in CITES will reverberate in the IWC.
>
> (2001a: 313)

Along the same lines, Iino and Goodman argue that the IWC's credibility of the IWC 'as a resource management organization' has been 'challenged by other international organizations'. As examples, they suggest that in opening statements at IWC meetings the IUCN has 'urged the IWC to complete the RMS' and the Secretary-General for CITES has stated in a letter that 'the IWC should soon make important progress towards the adoption of' the RMS (Iino and Goodman, 2003: 6).[49] The linkage between the two treaties is surely clear – both in the explicit steps taken by states parties, by international organisations and the treaties' secretariats to link them, and in the subject matter with which each deals and in the concerns they must share.

In the IWC, the tactic used by the anti-whaling members, of bringing in new members to support their views and eventually to adopt the 'moratorium', was adopted by the pro-whaling members – and the ICRW saw a period of rapid growth, with many states joining despite having no discernible interest in whaling. This 'numbers race' quickly led to deadlock. After various attempts by the pro-whaling states to overturn the moratorium proved fruitless, this impasse then led to efforts by these states to broaden the ambit of the debate and to involve CITES. A deliberate tactic was then apparently followed, by certain pro-whaling parties, of trying to circumvent the ICRW through downlisting motions at CITES. The ensuing years saw varied efforts being made to re-open commercial whaling, with each effort being stymied by the pro-whaling members. The relationship between the IWC and CITES became particularly dangerous ground as the years went by and as CITES was used by the pro-whaling states to bring increased pressure to bear on the ICRW. Other tactics used, and difficult relationships which arose, included the issues of small-type coastal whaling and aboriginal whaling, and efforts to transform the nature of the ICRW. Many efforts by well-meaning states to bring the warring parties together failed. Political manoeuvring dominates the IWC.

At the time of writing there seems to be little *immediate* prospect of bringing the impasse to an end. IWC members remain apparently locked into one camp or the other, with little room for middle ground. Parties who try to sit on the fence

49 The letter was written by then CITES Secretary-General, Willem Wijnstekers, on 4 July 2000.

risk alienating both friends and enemies. Most continue to complain and to express discontent with the situation. There is also a possibility, however, that many might in fact be relatively content with the status quo which means that there is even less incentive to find common ground.

3.13 More than mere diplomacy: Sea Shepherd and Japan's scientific research whaling

It is worth mentioning that the conflict in the Antarctic between the Japanese research whaling fleet and environmental protestors from the Sea Shepherd Conservation Society has been growing increasingly bitter in recent years. At the end of each year, Japan sends its fleet to the Antarctic, and Sea Shepherd (and occasionally Greenpeace) sends ships to observe, protest and, at least in Sea Shepherd's case, if possible prevent the taking of whales. The conflict between the Japanese vessels and Sea Shepherd is escalating.

From 2006 until present writing, the conflict has expanded rapidly from a single Sea Shepherd vessel harassing the Japanese vessels to numerous vessels equipped with helicopters engaging in a mini-naval war. There have been boardings of Japanese vessels, one of which resulted in a lengthy criminal court prosecution in Japan; there have been alleged acts of violence from both sides, including the hurling of objects onto Japanese vessels and the claimed shooting of Sea Shepherd Director Paul Watson (wearing a bullet-proof vest); there have been collisions between vessels and even the sinking of a Sea Shepherd vessel which was rammed by a Japanese vessel, whose fault this was still being in dispute; there have been attempts to foul propellers and disable vessels; and there have been apparent attempts by both sides to use acoustic or laser weaponry to interfere with the others. There have been court cases, with the most recent being a successful injunction against Sea Shepherd being obtained late in 2012 in a US District Court by the Japanese Institute for Cetacean Research and others, in which the court labelled Sea Shepherd as being guilty of 'piracy' – at the time of writing, Sea Shepherd had, however, declared its intention to appeal against the judgment. Numerous resolutions on 'Safety at Sea' concerning the conflict have been agreed to at the IWC, and Sea Shepherd, as an organisation, has been barred from attending IWC meetings.

It is impossible to know how this conflict might eventually be resolved. Sea Shepherd does appear to have been remarkably successful in curtailing the numbers of whales which the Japanese vessels have been able to take; has done much to win public sympathy in the West through this success; and has been able to raise significant funding from donors. The Japanese research effort has, however, been able to win new credibility both within and outside of the IWC by presenting itself as the victim of 'eco-pirates'. The result of the Australian/Japanese litigation in the ICJ may influence the matter in the future, giving new credibility to one side or the other, if there is a decisive judgment by the Court.

3.14 International litigation and scientific research whaling

Birnie writes that in 1946, 'in pursuance of the US proposals, a concession was made exempting from the regulations of the IWC whales taken and treated under scientific permits that Contracting Governments were allowed to issue for whatever numbers they thought fit' (1985: 190). The position remains that the IWC cannot prevent countries from issuing permits in their discretion. Of course, were it authoritatively determined that a party was 'abusing the privilege' – for instance, by taking far more whales than needed, or by taking them for reasons other than genuine research – the legal position might well be seen as different.

Japan's first research programme – JARPA (Japan Whale Research Programme under Special Permit in the Antarctic) – was an 18-year programme which ran from 1987/88 to 2004/5, and which had as stated objectives: the estimation of biological parameters (especially the natural mortality rate) to improve management; the elucidation of stock structure to improve management; the examination of the role of whales in the Antarctic ecosystem; and the examination of the effects of environmental changes on cetaceans. It involved the taking of 400 (with 10 per cent variability) minke whales annually (IWC 59, Chair's Report).

Japan's second Antarctic research programme – JARPA II – commenced with the first year of a two-year feasibility study in 2005/6. The objectives of JARPA II, as defined by Japan, are to monitor the Antarctic ecosystem; to model competition among whale species and develop future management objectives; to elucidate temporal and spatial changes in stock structure; and to improve the management procedure for Antarctic minke whale stocks. The focus is on Antarctic minke, humpback and fin whales, and possibly other species in the Antarctic ecosystem that are major predators of Antarctic krill (IWC 59, Chair's Report).

Japan also has the JARPN II programme – a long-term research programme aimed at studying feeding ecology in the context of contributing to the 'conservation and sustainable use of marine living resources in the Western North Pacific, especially within Japan's EEZ' (IWC 59, Chair's Report).[50] Japan's take in the North-West Pacific is a catch that does not ever seem to attract the attention of the NGOs who rail against the Antarctic catch – even though Japan takes sei whales, in respect of which it does not hold a CITES reservation (Sand, pers. comm., 2008, 2012).[51]

Norway, says Friedheim, claims that they set their quotas in accordance with the Revised Management Procedure (Friedheim, 2001a). On the one hand, setting quotas for scientific permits in terms of the RMP makes it more difficult to fault; on the other hand, it may well have been intended as a provocative act. In the world of the IWC, there are very often hidden motives and much manoeuvring for even minor advantage. Also, this move has the effect of conflating commercial

50 The programme involves the taking of Bryde's, minke, sei and sperm whales annually in the western North Pacific.

51 It has been reported that in the three month 'season' from June to August 2008, Japan caught 100 sei whales, 59 minke whales, 50 Bryde's whales and two sperm whales.

with scientific whaling – something that can hardly ever have been the object of the provision for states to issue permits for scientific research takes of whales.

It is uncertain to what extent such scientific whaling has benefited research. Lapointe argues that the:

> scientific data gathered by Japanese research vessels has provided the world with a sound scientific basis for a global whale management plan, a much needed conservation tool that has been stalled by Greenpeace and its colleagues for more than a decade.
>
> (2003: 65)

Gales *et al.*, however, wrote in 2005 that 'Japan's expanded programme will result in annual catches that are more than half the total cumulative catches for scientific research by all nations in the past half-century', and that 'such takes differ little in scale from commercial whaling, and must be justified by an adequate scientific rationale' (Gales *et al.*, 2005: 883–4). Japan does argue, however, that there are significant differences between commercial whaling and its scientific whaling.

At IWC 53, in 2001, in its (written) opening statement, Japan said that 'there has been a lot of misinformation concerning whales but perhaps the worst case of misinformation has been the criticism of Japan's whale research programs' and the labelling of these as 'illegal whaling' and 'commercial whaling in disguise'. Members of this Commission, the statement continued,

> should know that contrary to these claims, Japan's whale research programs are perfectly legal under Article VIII of the ICRW and that the Commission's Scientific Committee has commended both the quality and quantity of data from these programs noting that they are providing valuable information for the management of whale stocks.

Japan also pointed out that its 'whale research programs pose no risk to the whale populations' (IWC 53, Opening Statements). The claim of commendation by the Scientific Committee is put somewhat too strongly. The Scientific Committee, cautious by nature, has a tendency to suggest that Japanese research is useful, but not to endorse the claim that it is *necessary*.

Komatsu and Misaki suggest that the difference between research whaling and commercial whaling is that:

> Japan's Antarctic whale research has four components: these being to ascertain the biological parameter values necessary for the management of the minke whale stocks; to elucidate the role of whales in the Antarctic system; to elucidate the effects of the environmental changes to the whales, and to identify the differences of the stocks.
>
> (2003: 139)

The issue of trust provides a serious barrier to rapprochement. Anti-whaling states simply do not trust pro-whaling states, given their rapacious and dishonest behaviour in the past. It seems, for instance, that much of the official catch data submitted by the Soviet Union to the IWC in the 1960s and 1970s was false (Martin, 2001). A resolution 'on the Unreliability of Past Whaling Data' was even passed by the IWC in 1994 (IWC 46, Chair's Report). At IWC 47, 1995, the Chair's Report reflected the claim that the 'first report of massive under-reporting of Soviet pelagic whaling operations was made public by a Russian scientist in 1993'. The Russian delegation in 1995, however, argued that the data had not been submitted officially by the Russian government (IWC 47, Chair's Report). This is supported by Nikonorov, who claims that the former Russian commissioner and advisor to the President Alexey Yablokov was speaking 'on initiative of his own' (1996: 5).

Pro-whaling states do not trust the anti-whaling states either. Consider, for example, the words of the Japanese commissioner in his (written) opening statement to IWC 33, 1982, where he suggested, in respect of North Pacific sperm whale populations, that the previous year a new computer analysis had been used which had indicated that the male population was below the MSY level and that these figures had been used as the basis for reducing the quota. 'Subsequent to the meeting', the opening statement suggests,

> Japanese scientists were astonished to find that reiteration of the computer procedure as detailed by the description of the analysis did not produce the conclusion of the authors, and moreover that the analysis contained a number of serious errors in theory, computer programming and data handling. These results were communicated to and acknowledged by the authors.
>
> (IWC 33, Opening Statements)

Gambell writes that 'since its moratorium decision, the IWC has established detailed guidelines for its Scientific Committee to employ in reviewing and commenting on proposed special permits'; however, he goes on, 'critics of Article VIII continue to suggest that special permit catches are being used as mechanisms to circumvent the prohibition on commercial whaling until the moratorium is lifted' (2003: 80). Butterworth told me that in the 1990s the Japanese 'came on board with the science' and 'said we want to do it properly, maybe we haven't done it so well in the past, we'll try and do a better job of it'. However, in his view the Japanese have since 'come out of' that process, saying, cynically, that:

> it doesn't matter, we've tried to play ball on the science, it made no damn difference, look at what happens. On the RMP, we've done everything that was asked of us – the only thing that matters is how many votes there are.
>
> (Butterworth, pers. comm., 2006)

On the issue of whether scientific permit whaling is justified or not, Butterworth said that there 'certainly is policy involved in this as well' in the sense at least of keeping the industry functioning, even if at a low level – he described this as being 'a concern'. On the question, however, of whether scientific whaling is 'commercial whaling in disguise', he argued firmly that it is not. In this regard, he explained that the Antarctic whaling is subsidised, and that subsidies would not be needed 'if you could go down there and do it commercially'. According to him, the Japanese fleets are 'doing the whaling in a scientific way and they're covering the whole area'; whereas 'if you just wanted to catch you would go down to the places where the highest concentrations are', instead of having 'your fleet sailing up and down 300 miles' and going 'where the densities are very low' (Butterworth, pers. comm., 2006).

'The IWC', according to Burns and Wandesforde-Smith, 'has consistently called upon Japan and other parties that have conducted scientific whaling activities to desist, on the grounds that such research is not essential for rational management of stocks and does not address critical research needs', which requests, they add, have been 'rebuffed' by Japan (2002: 202–3). However, it is arguably somewhat disingenuous for the writers to claim that it is *the IWC* which has called upon Japan to stop. While technically correct, this overlooks the fact that such a 'call' would have been made simply by a majority (perhaps even a slim majority) of members in the divided organisation.

Iino and Goodman claim that both of the moratorium and the Southern Ocean Sanctuary apply only to commercial whaling, not to research whaling, and, therefore, that claims that Japan is violating either have no proper legal basis. They argue also that it is long-term research programmes which are required properly to assess ongoing changes 'of various population parameters since sampling for only one or two years does not tell you what is going on in a dynamic system'. Scientific research, they say, is critical to the IWC as Article V of the ICRW requires that regulations made by the IWC be 'based on scientific findings' (Iino and Goodman, 2003: 16, 18). Burke argues that the principle that the IWC must be guided by the best scientific evidence available ought to be given greater weight than it is currently given in implementation of the ICRW (2001). It appears to me, however, that there may (strongly arguably) be some 'abuse' of the 'ecosystem argument' by defenders of scientific whaling. It is clear that it is being linked to the idea that taking an ecosystem approach means finding that whales and humans are in conflict. It also arguably presupposes *maximised* taking from the sea. Payne *et al.* argue that much more than is generally realised can be learned from non-intrusive techniques, and that 'even when we had hundreds of thousands of corpses from which to derive data, the answers to some very simple questions (the consequences of which are nevertheless very important) were not forthcoming' (1983: 2).

From 2008, Japan's catch quotas are intended to see up to 1,415 whales taken annually – mostly from the Southern Ocean Sanctuary area. This compares with 273 whales having been taken in 1988. Sand records that, from 1988 to 2007, 10,857 whales were taken under the auspices of Japanese scientific permit whaling – compared to approximately 2,100 whales having been taken under the same auspices by all of the world's states in the period 1948 to 1987 (Sand, 2008). These figures provide probably the best basis for criticising the Japanese scientific whaling effort as illegal. Since scientific permit whaling is a right provided for in the ICRW, it is difficult, however, to describe it as a contravention of the moratorium (the 'zero quota on commercial catches') – and it is difficult even to describe a 'right' as a 'loophole.' To do so would require demonstration that the right is being abused and that it has therefore become a loophole, at best.[52] The Japanese response might even be that the situation has been forced upon them by the imposition of the zero quota in paragraph 10(e) of the Schedule to the ICRW – in other words, that increased scientific permit whaling is necessary in order to determine, through research, whether the zero quota remains necessary.

Scientific research into whaling, particularly into population statuses, has always been recognised as an essential feature of the work of the IWC. Since the coming into effect, in 1985/6, of the moratorium, however, whaling under scientific permit has taken on increased importance. It has become something of a mantra amongst anti-whaling parties, NGOs and media that Japanese research whaling is 'bogus' and designed rather to supply the market with whale meat. In the absence of an entity, the equivalent of an upper court in a national legal system, with the jurisdiction in international law to make such a finding, however, it remains impossible to state conclusively that the 'scientific permit exemption' is being abused. The 'jury is equally out' on whether the scientific research which the Japanese undertake represents genuine scientific research or not.

'Research whaling' is probably the biggest current bone of contention between the pro- and anti-whaling members. However, that Japan can continue to take substantial numbers of great whales annually despite the 'moratorium' may well be a factor contributing to a certain sense of stability – as long as it is able to use this 'loophole' Japan does not have a real incentive to leave the ICRW, a move which would potentially cause the treaty to collapse.

At the end of May 2010, Australia instituted proceedings against Japan in the ICJ. In grounding jurisdiction, Australia relied on its own declaration of 2002 recognising the ICJ's jurisdiction, and on Japan's similar declaration of 2007. Australia has alleged that:

> Japan's continued pursuit of a large scale programme of whaling under the Second Phase of its Japanese Whale Research Programme under Special Permit in the Antarctic ("JARPA II") is in breach of obligations assumed by Japan under the [ICRW], as well as its other international obligations for the preservation of marine mammals and marine environment.
>
> (ICJ, 2010)

In particular, Australia has contended that Japan is in breach of obligations under paragraph 10(e) of the Schedule in not observing in good faith the zero quota

52 Lars Walløe, Head of Science, Norwegian Delegation to the IWC, told me that the original intention behind the inclusion of the exemption for scientific permit whaling had been that only very small numbers of whales be taken (Walløe, pers. comm., 2007). The point was made to me off the record; however, the same point appears (attributed to Walløe) in Sand, 2008: 57 (quoting V. Morell, 2007).

catch limit, and is in breach of an obligation under paragraph 7(b) of the Schedule not to undertake commercial whaling in respect of fin and humpback whales in the Southern Ocean Sanctuary. Australia has contended further that Japan is in breach of its obligations under both CITES and the CBD of 1992. As remedies sought, Australia has asked essentially that the ICJ order that Japan cease implementing JARPA II, revoke any authorisations allowing activities which are the subject matter of the application and provide guarantees that it will take no further actions under JARPA II, or similar programme, until such time as such programme has been brought into conformity with its obligations under international law (ICJ, 2010). The ICJ then fixed dates in March 2011 for Australia and March 2012 for Japan, for the filing of written pleadings (ICJ, 2010). After delivery of both sets of pleadings the ICJ decided that no further pleadings (reply, rejoinder) were necessary, and that pleadings were now closed (ICJ, 2012).

In November 2012, New Zealand filed a declaration of intervention, which does not in fact make it a party to the litigation proceedings. New Zealand declared its intervention as necessary in order to place before the ICJ its interpretation of Article VIII of the ICRW. New Zealand's intervention can be summarised as being that in its view any special permit whaling is prohibited if not in conformity with Article VIII, and that Article VIII permits the killing of whales under special permit only if it can be demonstrated, by objective assessment of the methodology, design and characteristics of the relevant programme, that the killing is for purposes of scientific research only; that the killing is both necessary for and proportionate to the objectives of that research, and that it will have no adverse effect on the conservation of stocks; and that the duty of meaningful cooperation with the Scientific Committee of the IWC has been discharged by the contracting government which issues the special permit (ICJ, 2012). In February 2013, the ICJ decided that New Zealand's declaration was admissible, fixed a date in early April 2013 for the filing by New Zealand of written observations and authorised the filing by Australia and Japan of written observations, to be filed by the end of May 2013, on New Zealand's written observations (ICJ, 2012).

At the time of writing, the pleadings were not public documents, and my own impression from talking to Australian delegates at the IWC meetings in 2011 and 2012 is that the Australians are keeping their contents very close, while they do appear to believe that their case is a strong one. It seems most likely to me that the allegation that Japan is in breach of Article VIII of the ICRW will rely on a claim that the research conducted by Japan under scientific permit whaling is not genuine research; relying for this on various evidence, such as figures showing that the numbers of whales taken by Japan under scientific permit were insignificant until after the commencement of the zero quota at the beginning of 1986. Australia might also point to evidence, including the original treaty negotiations as well as practice up until the beginning of 1986, showing that the taking of significant numbers of whales under scientific permit whaling is not what was intended by Article VIII and is not therefore permitted by the ICRW. A significant difficulty for Australia will be that the IWC's Scientific Committee concluded in 2006 that data from the research 'provides a valuable resource to allow investigation of some aspects of the role of whales within the marine ecosystem' (quoted in Goodman, 2010: 179).

In respect of the allegation that Japan is in breach of CITES, it is possible that Australia might argue fairly straightforwardly that Japan is acting in contravention of CITES by engaging in trade in Appendix I-listed species, but this would be a difficult argument to sustain given that Japan holds reservations to the CITES listings. My guess, instead, is that Australia's argument here will be based upon the prohibition in Article III.5 of 'introduction from the sea'[53] of species listed on Appendix I, without the prior grant of a certificate from Japan's management authority – which certificate may only be granted if a scientific authority of Japan has advised that the introduction will not be detrimental to the survival of the species involved, and the satisfaction of a management authority of Japan that the specimen introduced will not be used for primarily commercial purposes. While Japan's national authorities would probably certify such advice and satisfaction, Australia might make the allegation that such belief is either not genuine or not reasonable. At only the second CITES CoP, in 1979, a resolution was passed in terms of which parties agreed not to issue for primarily commercial purposes any import or export permit, or certificate for introduction from the sea, for any specimen of a species or stock protected from commercial whaling by the IWC. It might be that Australia will be relying on this resolution, which was in fact approved before Japan joined CITES but the essence of which has been confirmed since (for instance at CoP 12 in 2002). However, once again, the reservations which Japan holds might be a barrier which Australia is unable to scale.

In respect of the allegation that Japan is in breach of the CBD, my guess is that Australia will rely particularly on the first substantive principle (Article 3) which is to the effect that while states have the sovereign right to exploit their own resources pursuant to their own environmental policies, in the environments of other states or in areas beyond national jurisdiction such states have a responsibility to ensure that their activities do not cause damage. This principle began its life in the Stockholm Declaration at the UNCHE in 1972, and it has been repeated in so

53 Between 2010 and 2013, the CITES worked on a process of clarification of this phrase (which is unique to the CITES) for consideration at CoP 16. The draft text recommended that where any species listed in Appendix I or II is taken in the marine environment not under the jurisdiction of any state by a vessel registered to a state, and is transported into that state, that state shall be the 'state of introduction'; and where such specimen is taken by a vessel registered to one state and transported into another state, those states shall be the states of 'export' and 'import' respectively. The draft recommended that states of 'introduction', 'export' and 'import' should (prior to issuing relevant permits) take into account whether or not the specimen was or will be acquired and landed 'in a manner consistent with applicable measures under international law for the conservation and management of living marine resources, including those of any other treaty, convention or agreement with conservation and management measures for the marine species in question'; and whether acquired and landed 'through any illegal, unreported or unregulated (IUU) fishing activity'. In Committee II, a vote ran 56 for, 15 against, with 14 abstentions – all of Australia, Iceland, Japan, New Zealand and Norway voted for. The text was adopted in plenary (CITES, 2013).

many subsequent international instruments that it is now strongly arguable that it has become a principle of customary international law. The difficulty for Australia will be that of proving that Japan's activities are in fact causing damage. Australia might also rely on Article 5, which provides for a duty on states party to cooperate with other states party, either directly or through competent international organisations, for the conservation and sustainable use of biological diversity[54] in areas beyond national jurisdiction. The difficulty here will be twofold: to show that Japan is not cooperating, and to show that whaling under scientific permit is not in the interest of conservation and sustainable use.

I am surprised that Australia commenced proceedings in the ICJ. It is a risky move. A clear-cut declaration by the Court that Japanese scientific permit whaling is in breach of international law would certainly be of huge value to the anti-whaling side's arguments; but, likewise, a decision favouring Japan's arguments would make it extremely difficult for the anti-whaling countries to maintain the argument that Japan is acting illegally. And the nature of the ICJ is such that a decisive, unanimous answer can probably not be expected. It is a forum in which a litigant state is entitled to choose an ad hoc judge if one of the judges already serving a term on the bench is not of its nationality, which shows something of the difference in the nature of the Court when compared to a national court. It is a forum which will also be grappling not with the interpretation of firm statutes promulgated by designated authorities as in national legal systems, but with the vagaries of international law in a unique situation where the primary 'statute' governing the matter does not itself provide for an arbiter or any other means of resolving disputes. An indeterminate decision, or a decision with dissensions or separate differing opinions, would be more to Japan's advantage than to Australia's. As such, while the eventual decision does have the potential to change IWC debates radically, I think that it most probably will not do so.

54 'Sustainable use' is defined in the CBD as the use of components of biological diversity in a way and at a rate that does not lead to the long-term decline of biological diversity, thereby maintaining its potential to meet the needs and aspirations of present and future generations. 'Conservation' is not defined.

4 The second old watchdog

CITES

4.1 Creation

CITES is an 'old-style' Convention which does not make provision for ongoing monitoring (although it has a permanent secretariat charged with ensuring that it is operating successfully, day-to-day operation is a matter for the national authorities of the parties) and which does not make provision for financial support of countries which have difficulty in meeting their treaty obligations due to lack of funds, or which are forced to act under the treaty at undue financial cost to themselves relative to other parties. It was signed into existence at Washington in 1973, as one of a number of conventions surrounding the 1972 United Nations Conference on the Human Environment (UNHCE) in Stockholm. CITES regulates international trade and has only extremely limited (and largely informal) jurisdiction to look at situations inside states and to regulate their internal actions.

On the history of CITES, it has been suggested that it may have had its early origins in the US – in the 'Lacey Act of 25 May 1900, which prohibited interstate commerce in illegally taken wildlife', and which was 'extended in 1935 to wildlife imported from abroad' (Sand, 1997: 20). Of course, it is just as likely that CITES's origins are to be found in the 1900 London Convention (influential, although never coming into force) and the 1902 Paris Convention – and the ideas of categorisation of species which flowed from these two early conventions. In truth, the origins of CITES must be seen as lying in many places – and indeed in the *zeitgeist* of increased environmental awareness and concern in the late 1960s and early 1970s.

In the US, the Endangered Species Conservation Act (amending the ESPA of 1966) of 1969 authorised the US Department of the Interior to promulgate a list of wildlife 'threatened with worldwide extinction', and to prohibit imports thereof except for scientific or breeding purposes. At the same time, records Sand, the US government was 'directed to encourage' similar laws in other countries and to 'seek the convening of an international ministerial meeting' to conclude 'a binding international convention on the conservation of endangered species'. This initiative 'coincided with' the UNCHE, as well as with research within the IUCN which had, in 1963 at its General Assembly in Nairobi, called for an 'international convention on regulation of export, transit and import of rare or threatened wildlife species or their skins and trophies'. The IUCN suggestions had been based on the idea of controlling (or banning) the wildlife trade 'on the basis of global lists of threatened species to be drawn up and updated (along Red Data Book lines) upon advice by an international expert committee'. Opposition to this approach, according to Sand, came from developing countries 'led by Kenya, insisting on the right of each range state to determine its own list of tradable species'. Nevertheless, both approaches were ultimately accommodated in a 1972 US draft that served as working document for the conference of 80 potential parties in 1973. The outcome was CITES which was both a 'conservation *and* trade instrument', to protect wild fauna and flora both for humankind ('present and future generations') *and* as national heritage ('of peoples and states') (Sand, 1997: 20).

According to the CITES Secretariat itself, CITES is a treaty for preventing over-exploitation of wild fauna and flora for international trade (CITES, 2006). Supported by national legislation, according to the CITES Secretariat, trade in species listed in the three CITES Appendices must conform to CITES requirements and provisions outlined in the text of the Convention. Ongoing implementation of CITES is steered by resolutions of the CoPs. Permits, in standardised formats, are issued by competent authorities where certain conditions in the exporting country, or in both the exporting and importing countries, are met. Conditions vary according to the Appendix in which particular species are listed, focusing on the purpose of trade, the impact on conservation (as determined by a scientific authority), legal acquisition and, if relevant, humane treatment concerns. Permits are endorsed (usually by Customs) upon exit, and presented (usually to Customs) on entry. Data collected from issued permits contributes to a body of information that allows parties to follow international trade trends, and supposedly then to adapt national and international conservation and trade policies as necessary.

4.2 Nature

CITES does appear on the whole to have been a successful treaty. Harland calls it 'the most active and visible instrument of international environmental law in force' (1994: 12), and Hepworth argues that most parties believe CITES works 'as it has direct, practical effects in the real world' (1997). Perhaps chance has played a role here, in CITES having a large membership. Perhaps the timing, being some 25 years later than the ICRW, led to the greater membership and to greater commitment by members. Perhaps it is that the ICRW remains basically a single-issue treaty, in a world where greater understanding of complexity and inter-relationships (in nature, and in man's relationships with nature) have come to dominate thinking. Both treaties arguably have a dual purpose (trade, or use, and protection) but CITES is certainly more specifically protective. Or perhaps flexibility is the key, and perhaps, then, this should be built into treaties, in recognition of inevitable self-serving of interests.

However, while it may be true that CITES is the 'most active and visible' of international wildlife-related treaties, and perhaps even the best-known of all international environmental legal treaties, it is not necessarily the best understood.

It is strongly arguable that CITES is not always well-understood 'on the ground' (in other words, within national constituencies) – even by persons actively involved in decision-making that affects the international operation of CITES.

Initial growth in CITES membership was slow. At the time of writing, however, membership stands at 178 – more than double the list of original signatories – and momentum was clearly gained. Neither CITES nor the ICRW restrict membership. An amendment was even adopted in order to allow the European Economic Community to become a party to CITES. In fact, there are no major trading states which are not party to CITES – with the exception of Taiwan. Taiwan's is a significant omission, as it is a major trading force, but its problematic relationship with China, and that country's influence, has thus far kept it out. The ICRW (88 parties at the time of writing) has a membership half of that of CITES, but ICRW membership is generally growing (despite one withdrawal in 2013).

This philosophy of inclusiveness can arguably be seen as a weakness in CITES. Lapointe argues that it is a problem that NGOs 'never totally left the arena, nor dropped that sector of influence where they had total hegemony' and that this has resulted in a 'dual leadership: one official and one underground, and, in most cases, both are totally irresponsible and incompetent, because they have agendas and objectives irrelevant to conservation' (2003: 21, 48). On the other hand, it has been argued that cooperation with IGOs and NGOs – particularly the IUCN, the World Wildlife Fund (WWF) and the TRAFFIC network, although CITES has many linkages – has given CITES both a high degree of transparency and significant operational information sources (Sand, 1997).

CITES is often described as being both a protectionist and a trade treaty.

Reeve argues that CITES is not:

> per se a treaty to promote trade and use of wildlife. The principal objective of CITES is and has always been to ensure that international trade does not lead to species extinction. In other words, CITES seeks to prevent unsustainable use, not to promote sustainable use over non-use.
>
> (2002: 28–9)

This assessment seems to imply that protection is the objective more than is facilitation of trade – or at least that the former slightly outweighs the latter. This could be important insofar as it might suggest that too much protection, or at least a perception thereof, could explain state reluctance fully to embrace the philosophy of and obligations of CITES – while membership, and ratifications, of the treaty are high, enforcement is not as strong as might have been hoped for. David Western, though, feels that CITES is not a conservation treaty but a trade treaty. In his view, the fact that CITES says nothing about appropriateness of measures inside countries shows that it really has to do with trade. He says that CITES is 'so singular in its approach that it doesn't have the sophistication to deal with more than trade' (Western, pers. comm., 2004).

It seems to me that the core difficulty which CITES may be overlooked, however. In the light of increasing understanding of the ways in which species interact with each other within ecosystems, and that it is often habitats that need to be protected rather than individual species, by continuing to take an approach of categorising species, CITES is not in line with present needs. It is often said that CITES suffers from difficulties caused by its having two 'contradictory imperatives' – protection and trade. It might be, however, that trying to reconcile protection of species (as an aim) with categorisation of species (as a modus operandi) provides an even more fundamental difficulty.

On enforcement, Reeve suggests further that CITES is one of the oldest MEAs and that it is a 'specific tool rather than a global solution' (2002: 6). Probably what she means by this is that, like older MEAs, it concentrates on a particular issue rather than on a holistic approach – the latter approach being what the CBD arguably tries to take. On perceptions of CITES, however, Reeve suggests that CITES has, through its lifetime, been seen as 'the flagship wildlife agreement'. It is, she argues, a 'cornerstone MEA' which is at 'the nexus of international trade and species conservation' (ibid.). Further, while there are suggestions that many perceive CITES as being effective, it can clearly only be said to be generally effective. On a daily basis, it is breached. Or perhaps it can be said to be effective, but only for its (limited) purpose.

Reeve does suggest that by operating through resolutions and decisions, through 'soft' law, the CITES regime has become 'dynamic and flexible' and able to 'circumvent the delay imposed by treaty amendments' (2002: 41–2). One possible advantage of 'soft' law might well be that it enables participants to 'feel' their way toward what the law *ought* to be before it becomes binding. In this way, when international law becomes binding custom it is tried and tested as a true reflection of actors' true beliefs and understandings. There are, however, concerns about CITES. Perhaps one thing that CITES lacks, and has lacked since its inception, it occurs to me, is a true incentive, or system of incentives, for states to embrace its requirements fully. According to the CITES Secretariat,

> as a decision-maker in the trade chain, the public is a major stakeholder in the regulatory process. If the public does not understand or support the Convention, this makes the task of implementing CITES much more difficult and this, in turn, affects the effectiveness of the Convention.
>
> (CITES, 2006)

In more than just the matter of national implementation, many states pay lip-service to their CITES obligations. This is a serious problem. Without effective implementation, a treaty cannot work.

Reeve explains that CITES is a 'non-self-executing treaty' in that national legislation is required to implement its provisions. Parties must take measures to prohibit trade in specimens violating the Convention, and to provide for penalties for violations and for the confiscation of specimens (Reeve, 2002: 52). While it is easy to castigate states for not enforcing provisions of treaties to which they have become party, there might, of course, be many reasons why such enforcement does not happen.

Always, there is a watering down of meaningful obligations in MEAs. An extensive revision of the non-compliance procedure might improve the treaty's effectiveness, but would need to be carefully handled. Given the history of CITES, the natural temptation would be to focus on giving more 'bite' to trade sanctions. However, these might not be the best form of sanction. Apart from possibly bringing CITES into conflict, thus far avoided, with World Trade Organization (WTO) rules, trade sanctions might be effective (especially as achieving the object of CITES) but would carry the inherent danger of implying recognition of CITES as a trade treaty – and perhaps even overemphasising this aspect at the expense of the conservation aspect.

Enforcement is a thorny issue for both CITES and the IWC, with the IWC at times in its history facing even the more serious problem of falsified returns by states instead of the absence of returns. Perhaps, then, the US has threatened to resort to a form of unilateral trade suspensions to punish what it sees as breaches of the ICRW because of the absence of the CITES option of formal trade suspensions as a sanctioning mechanism. What this implies is that it is trade that is used as punishment, reinforcing the trade-related nature of the two treaties.

This trade-related nature is emphasised by Ong, who writes that although CITES may have been successful in providing for strict controls on trade in endangered species, it is 'at least arguable' that its stated aim 'in terms of the protection of endangered species against over-exploitation though international trade controls implicitly acts to legitimise such trade in fact' (1998: 294). Pickover reinforces this perception, suggesting that a negative aspect of labelling certain species as 'rare' is that it 'adds market value and increases demand', with many species declining in spite of their Appendix I listing because of relentless market demands (2005: 55).

CITES is not designed in itself to protect or conserve endangered species, with the possible exception of species threatened with extinction and listed in Appendix I. Appendix I species are not available for commercial trade and are thus ostensibly protected against trade. They do, however, face other dangers. CITES functions through a permitting system which has in the past proved problematic and which did little – if anything – to stem the ivory trade in the 1980s. In respect of ivory at least, the discredited permit system has been done away with.

It probably is wrong to see CITES as a general wildlife management treaty – whether it ought to be such being a different discussion. As it stands, according to Sand, CITES is:

> but one component of the existing patchwork of global and regional wildlife regimes, narrowly focused on the transnational trade issue, which is only one of the multiple threats to wildlife; and hence should be judged by its contribution to mitigating that particular threat.

He argues that CITES has a broad role to play since, unlike more specialised 'unit management regimes' (such as the IWC), CITES does not control the 'actual taking of wildlife' – through a moratorium or catch quotas or prescribed methods of capture (Sand, 1997: 26).

4.3 Effectiveness

The increasingly influential position of developing countries within the CITES decision-making process can be seen as an example of democratisation of international environmental treaty regimes, as such treaties seek to accommodate many different perspectives and value orientations (Ong, 1998). Developing countries are increasingly aware of the bargaining power which they hold in respect of their biodiversity and this awareness is reflected in more modern treaties. However, one must not overemphasise this point – it is still the developed countries which have most bargaining power in international fora. The subtle threat to destroy, or not to take sufficient care of, one's own natural resources can never be a particularly strong bargaining chip. In a rapidly changing world, of course, it was never going to be easy to create a treaty that could accommodate all interests. Treaties – MEAs – are of necessity changing.

A number of commentators feel that CITES never did have a chance of working as effectively as intended. Parker writes that decline is *not* per se the outcome of trade and that there are many cases where trade has not reduced populations with there being other reasons for decline, such as changed habitat. In what circumstances, he asks, *does* trade diminish wildlife? The crucial factor, he argues, 'seems to be the absence of a sense of ownership' with unsustainable harvesting happening most commonly with wildlife in respect of which there is no sense of ownership (Parker, 2004: 313–5). Parker's argument is, of course, the 'sustainable use' argument – that trade is not a bad thing, and might even be a good thing.

A rather different criticism to Parker's comes from Pickover, who writes that 'CITES has failed wild animals because it does not provide for issues of ethics and compassion' and that it 'remains silent on the basic ethical question of whether it is even appropriate to engage in international trade'. Further, she argues that CITES appears 'concerned not so much with protecting species as with allowing trade in endangered ones, which makes it almost Orwellian in character' (Pickover, 2005: 54). This is the argument that CITES is too much in favour of the 'sustainable use' philosophy. The argument might be correct, but more probably it is correct only for certain times and in certain areas. The wording of the objects clauses of the treaty is strongly rooted in protectionism – which makes sense given that the treaty's negotiations began in the late 1960s and early 1970s – whereas the prevailing *zeitgeist* in the early twenty-first century is more that of sustainable use.

Sand argues that CITES has a number of problems – and even that it 'may indeed have reached its outer limits'. The problems which Sand points to concern the influence of free trade areas with concurrent abolition of internal trade boundaries; problems inherent in the inability to regulate trade in species that fall outside the treaty's ambit; and problems caused by competition from regional regulatory regimes (Sand, 1997: 26–7).

Parker writes that the ineffectiveness of conservation measures in Africa has more to do with lack of government will and funding than with the law itself. He suggests that if laws fail because authorities are too under-staffed or under-funded to apply them, then the solution is adequate staffing and funding, rather than new laws.

Parker explains that the 'manpower CITES needed for it to work' has never been estimated, and that 'conservationists rushed to list species' (2004: 315–6). On numbers of species listed, in 2013 it appears that there are roughly 5,000 species of fauna and 29,000–30,000 species of flora protected under the three appendices (CITES, 2013).

Defending the permit system in relation to the particular issue of ivory, Parker makes the point that although CITES was 'ineffective' it nevertheless 'made political sense for legitimate international ivory merchants to try to help it' (2004: 318). Arguing in favour of trade being necessary for successful conservation, Parker's line is that the ivory trade could have been controlled successfully and conservation thereby supported. Parker then argues that CITES is not, in fact, a conservation treaty at all, as it 'was used by the conservation lobbies to stop trade' and 'any steps to streamline and improve ivory commerce would be opposed no matter how well founded the arguments'. 'Perhaps the most cogent comment on CITES', Parker continues, comes from actions taken by the US and the EU, both of which 'enacted national laws that override CITES and render the treaty obsolete'. If CITES was workable, he concludes, 'there would have been no need for such additional national rules' (ibid.: 321–3). A point that Parker misses, however, is that CITES actually states explicitly that states may take stronger action than does CITES (Article XIV). It is hard, therefore, to agree with his assessment that the treaty has been 'rendered obsolete', even if he is speaking only in the context of trade in ivory.

Lapointe argues that the 'international community must realize that banning trade and closing legal markets is not progressive conservation, but an anachronistic approach that does more harm than good' and that 'developed nations have benefited from the use of their resources and are wealthier and healthier as a result. There is no ethical justification to deny people of less developed countries those same opportunities' (2003: 125–6). Arguably, however, there *is* such an ethical justification – that regardless of the rightness or wrongness of denying developing countries the same opportunity as was used by developed countries, it remains ethically wrong to use natural resources to the extent that the developed countries used them.

'More and more of the governments represented at CITES', continues Lapointe, 'have become aware that they have a national interest in paying attention to every item on the CITES agenda', concluding that such governments are coming to 'understand that if they fail to support other governments that pursue sustainable use policies, they are weakening the entire concept of sustainable use' (Lapointe, 2003: 125–6). This is an argument that can be used to justify, for instance, Israel's or Mongolia's or Senegal's accession to the ICRW. It cannot be denied that, as much as critics might argue that such joinders are merely opportunistic acts where the states involved would seem to have no direct interests, there *is* potentially important linkage.

It may be that despite, or perhaps because of, its visibility, the importance of CITES is overstated. Nevertheless, it cannot be denied that – whether there is one battleground or several – what happens at the high profile conservation fora provides crucial precedent. It is not because of the very importance of these fora that countries find it so difficult to agree within them; if the arguments were not of vital importance, compromise and resolution would be far easier to come by.

The charge, often made, of lack of political will is problematic. It may well be that governments which act against their international obligations to increase environmental protection are, in fact, following the political will of their constituencies – many of which will not prioritise environmental protection over more immediate concerns. Countries have pressing domestic economic, political and social problems, and politicians usually have relatively short careers.

Historically, the Secretariat of CITES has played an active role – even going so far as to provide its views, and give its personal opinions, on controversial proposals (Reeve, 2002). Former Secretary-General of CITES, Willem Wijnstekers, has written that:

> things would be a lot easier if the public realized that live specimens are not exempted from CITES provisions, that taking specimens of Appendix-I species home is not allowed, but that there is no problem where non-commercial quantities of Appendix-II species are concerned.
>
> (Wijnstekers, 2007)

While hedged with caution, assessments such as these clearly show that sustainable use is the most viable future path – at least in the Secretariat's view. This is, of course, the direction in which the majority of the world's states appear to be heading.

The Standing Committee of CITES has become an influential decision-making body, but has been controversial.[1] Despite assertiveness by the Secretariat and the importance of certain committees, Lapointe's assessment is that 'CITES has never realized a fraction of its potential'; but his conclusion is far too simplistic when he argues that this alleged failure 'can be attributed to the simple fact that, in many instances, parties to the Convention have forgotten the principles laid down in the treaty's preamble' (2003: 126–7). In other words, the argument is that had CITES

1 Reeve explains that there are guidelines for the duties of regional representatives. These, she says, 'fall short of an overtly expressed duty to *represent* the opinions of other parties in the region at Committee meetings, [but] it can be argued that it is implicit'. She then gives an example where such duty was not fulfilled. Her example is SC41 in February 1999, when ivory sales (arising from CoP 10) were approved. Three of the four parties wanting to trade ivory (Japan, Namibia and Zimbabwe) were, she says,

> members of the Standing Committee: Japan and Namibia were regional representatives with votes, while Zimbabwe was there by virtue of its role as host of COP10. However, instead of representing its region, Japan represented Japan – an ivory consumer – not even paying lip service to India's strongly held position opposing the trade. Namibia represented Namibia and the ivory trade lobby, and though several Southern African states supported the ivory sales, not all did, Zambia being a case in point.
>
> (Reeve, 2002: 267–8)

adhered to its ('pro-trade', according to Lapointe) purpose, instead of becoming bogged down in conflict, the parties could have made better use of its scope for flexibility and become a truly effective instrument. Sand agrees that CITES has inherent flexibility; in fact, that it was deliberately designed to be adaptable to changing circumstances (Sand, 1997).

Criticism for CITES comes from Leakey, too, at least in respect of the ivory trade. He argues, once again, that the problem was enforcement, with the organisation having no way to enforce its own rules or impose fines on those who broke them. He concludes, cynically, that:

> the more I found out about CITES and the ivory trade, the less sure I became that there was any way to manage it legally. As long as ivory was considered a valuable commodity, there would be a black market for it. Despite CITES and its well-intentioned rules, the ivory trade was clearly corrupt, and no matter who regulated it or what system they put in place, it would always be corrupt. The high price assured that.
>
> (Leakey and Morell, 2001: 42)

This, of course, describes CITES as it was up until 1989. In respect of the ivory trade, Leakey might well now agree that a CITES Appendix I listing can be fairly effective – the legal ivory trade having slowed considerably after the 1989 CITES Appendix I listing, for which Leakey had campaigned. The potential problems, however, remain.

Bonner tells us that 'most conservationists would argue that CITES has failed to live up to its noble intentions' and argues that, since the signatories gave the Secretariat no enforcement powers, if the treaty's 'goals are to be realized, individual nations must enforce its provisions. Most have lacked either the ability or the desire – or both – to do so' (1993: 95). Bonner therefore agrees with both Reeve and Leakey on the enforcement problem within CITES, and impliedly suggests that Parker is not correct in arguing that the treaty has been 'rendered obsolete' by the unilateral actions of states – in fact, quite the opposite.

The suggestion has been made, by Lapointe, that one of the major problems of CITES is the involvement – and undue influence – of NGOs. He argues that CITES CoP 7 in Switzerland, 1989, is 'by far the best example of the total domination of non-elected bodies over sovereign states represented within the CITES community' (Lapointe, 2003: 127–8).[2] The 'total domination' referred to concerned the decision to list the African elephant on Appendix I – it is certainly true that the 1989 CITES CoP saw a significant public relations effort by NGOs reach a successful climax; however, to argue that they were wholly responsible, or that they have always exercised so important an influence, would be to claim too much.

2 At this CoP, Lapointe, then Secretary-General, received considerable criticism from various NGOs, especially for acceptance by the Secretariat of donations from ivory traders.

According to Reeve, efforts to improve CITES's enforcement mechanisms have been made, but have not been successful. She argues that it does not appear as though such attempts are still being seriously made and she ascribes this (for reasons that are not clear) largely to the role of the Secretariat. She does concede that this does not mean that the Secretariat is unconcerned about enforcement (Reeve, 2002: 225–7). Kreuter and Simmons comment on the lack of enforcement ability within CITES, arguing that it might not be because of the Appendix I listing that the ban on the ivory trade has been – by and large – successful. They argue that 'it is important to recognize that the apparent success of the ivory-trade ban in closing most international markets and effective elephant conservation are not synonymous' (Kreuter and Simmons, 1994: 47–9). This takes one back to the criticism of CITES that the treaty might in fact stimulate trade.

It has been suggested that the CITES resolutions, which over the years have dealt with the organisation's relationship to the IWC, have all been based on the assumption that there is 'conformity between the agendas of CITES and the IWC, and that IWC policy is based on conservation, biological criteria and scientific advice – not whale rights and public sentiment' (HNA, 1997). However, CITES was later, in the 1980s, to assume 'an independent position in relation to the IWC by listing whale populations for which the IWC was still setting commercial quotas in Appendix I. This applied to the populations of sei whales, fin whales and Bryde's whales' (ibid.). Such independence could, arguably, 'be said to have implied indirect criticism of the IWC for allowing the commercial harvest of populations whose statues (at the time) was uncertain' (ibid.).

Thus, there are ironies. One is that CITES was once more protective than the IWC, and so parties within it may have tried to 'take over' the role of the IWC – but today CITES is being used by the parties who had, then, kept the IWC running as a use-oriented organisation, to undermine the IWC today. It is one of the contentions of this book that it is currently impossible fully to understand either treaty in isolation from the other.

5 CITES and elephants

5.1 East Africa, ivory and the 1989 CITES CoP

In the early 1970s, there was little support for a ban on trade in elephant ivory. I and O Douglas-Hamilton write that the conventional view of elephant conservation at the time was that 'it was more important to win the co-operation of the ivory trade rather than try and close it down' (1992: 29). By the mid-1970s, however, it was becoming apparent that elephants were in increasing trouble. Kenya, however, then become the first range state to take steps to control the ivory trade when, in December 1977, President Kenyatta announced a ban on all trade in ivory. I and O Douglas-Hamilton describe Kenyatta's decree as having been 'a revelation', and query what might have made him take that decision; speculating that it might have been that the price of coffee had soared, providing more lucrative returns than ivory for the moment. Alternatively, they speculate further, it might have been the power of the media, the word of the scientists or discreet persuasion through diplomatic channels that finally made some impression; but they advise that we will probably never know (I and O Douglas-Hamilton, 1992). In any case, from a morass of competing interests, a range state took a decision which was not apparently in its own immediate interest – and much of the rest of the world eventually followed suit.

Parker writes of Kenya that 'the hunting ban in 1977[1] and the revocation of all dealer's permits in 1978 were supposed to have ended all hunting, capturing and trading wild animals or their parts and derivatives'. Kenya, Parker cynically explains, had formally acceded to CITES in 1979, but the country had 'paid lip-service' to its rules 'since its inception' (2004: 187). Yet today Kenya is a leading protagonist in CITES. So, like many such, Kenya's current 'preservationist' stance may have its origins in cynical motives.

According to Meredith, many people prominent in conservation circles still argued that the greatest threat facing elephants in East Africa 'was overpopulation in protected areas for which the solution was culling'. The question was asked whether elephants should be allowed to find their own level 'in relentless

1 Most commentators refer to this as a 'ban' – however, Western told me that it was, and remains, a 'suspension' (pers. comm., 2004).

competition with other species' for limited resources, or whether they should be 'brought into balance with the environment and other species that inhabit it?' (Meredith, 2001: 204–5). Major disagreements as to the course which should be followed were beginning to appear amongst some of the chief protagonists. Loosely put, these protagonists comprised those who felt that it was necessary to control the trade while reaping its benefits – and that such control was possible – and those who were beginning to feel that a complete ban was what was required. It later became clear that the permit system simply was not working – although it had its defenders.

Meredith suggests that in 1979 two major reports were published, both of which were 'swiftly engulfed in controversy'. Iain Douglas-Hamilton, Meredith says, 'called for measures both to reinforce protected areas and to tackle the burgeoning illegal ivory trade', while Ian Parker 'provided a mass of new detail about the workings of the ivory trade', confirming that 'the trade had reached levels not seen since the early twentieth century' and calculating that 'between 1973 and 1978 about 1 000 tonnes of ivory a year had left Africa'. However, Parker apparently concluded that the ivory trade was 'not excessive', and in fact that it 'fell well within sustainable limits; and that it did not threaten the elephant's survival'. If anything, Parker allegedly said, 'the threat came from population growth and loss of habitat', and he 'went on to accuse conservationists of using false statistics to conjure up an elephant crisis to enable them to raise funds for their own purposes' (Meredith, 2001: 206–9). This represents, therefore, an early accusation of hypocrisy by 'conservationist' ('preservationist') NGOs – a charge which is more regularly made today.

It seems with hindsight that Douglas-Hamilton's figures were probably the more accurate, and that elephants were in serious danger from increasing poaching. Meredith explains that Parker's stance was 'fundamentally flawed', and that he 'had underestimated the real impact of the ivory trade by failing to take account of the way in which an ever-increasing toll on elephant populations was needed to sustain the same volume of trade' (Meredith, 2001: 206–9). At the time, from 1985, the system was that each country with elephants would annually declare an ivory export quota, and that every tusk leaving Africa would require an export permit, with the issuance of these permits being controlled by the producer states, in collaboration with the CITES Secretariat (Poole, 1996). Each quota was to be based on a domestic management programme for the sustainable utilisation of elephant stocks. Exporting states were 'required to notify the CITES Secretariat of any ivory consignment exported under its quota and to mark each tusk indelibly to identify its quota number', and 'importing states were required to accept only ivory shipments covered by valid documents' (Meredith, 2001: 211–4).

This, of course, is how the quota system was *supposed* to work. Meredith's judgement is that it was a 'total failure'. He ascribes this to the influence of 'legions of armed gangs, corrupt officials, greedy politicians and rich traders at work in the trade' (Meredith, 2001: 211–4). The CITES Secretariat was unable to enforce the rules, and it effectively accepted money from the fox for the guarding of the henhouse – its own 'ivory unit', writes Meredith, was so underfunded that to keep working it required 'donations' from wealthy ivory merchants. Between 1985 and 1989, he writes, 'two-thirds of its budget was provided by ivory dealers' (ibid.). The belief of the CITES Secretariat, however, was that the quota system would work given sufficient time, and that to work through a ban would be to risk driving the trade underground (ibid.). Much the same argument is made today, that poaching will not cease until the legal trade is reinstated and functioning efficiently. In fact, this argument is arguably the driving force behind CITES as a trade treaty – perhaps even its *raison d'être*.

Apparently in response to 'international pressure', in 1986 Japan imposed strict controls on ivory imports (Jackson, 1990). The slaughter, however, raged on. Leakey writes that elephant poaching had become 'nearly a continent-wide plague' (Leakey and Morell, 2001: 36). In some countries, such as Uganda and Sudan, he says, where there had been more than 100,000 elephants in the 1970s, it would have been difficult ten years later to find 10,000. In Tanzania, he records, where 100,000 elephants had lived in the Selous Game Reserve in 1976, a decade later, that figure had dropped to 55,000. Three years later, in 1989, the figure was 27,000. In Kenya, according to Leakey, 'we were down to fewer than 20 000 elephants – and still losing on average three a day'. It didn't take a genius, he says, 'to figure out that such a high rate of killing couldn't go on indefinitely' and that 'the African elephant was clearly headed for extinction throughout much of its range' (ibid.).

It was not only in East Africa that poaching was rife. Parker suggests that he 'knew of only three countries where ivory was openly and properly accounted for: Botswana, Malawi and Zimbabwe' (2004: 269). Even when states apparently took steps to halt the slaughter and the trade, these efforts were strongly hedged with self-interest. According to Reeve, in 1989, the NGO EIA claimed that over 1,000 tonnes of illegally obtained ivory had passed through Dubai since 1984 – where the official figures released from Dubai reflected merely that the UAE had imported only some 42 tonnes of 'unworked or simply prepared ivory' in 1987, with the figures being only 24 tonnes in 1988 and less than 2 tonnes in 1989. The UAE, according to Reeve, 'had actively facilitated the ivory trade by notifying its withdrawal from CITES just one month after traders from Hong Kong had been issued licences to set up further factories' (Reeve, 2002: 104–6).

Actions were, however, beginning to be taken. Reeve tells us that in 1986 'unilateral action was taken by the United States under the Lacey Act'.[2] In September 1986, the US Fish and Wildlife Service (USFWS) banned all wildlife imports from Singapore, citing the country's inability to provide 'comparable documentation' (Reeve, 2002: 129–30). The US was therefore acting unilaterally in respect of elephants, which it had really only *threatened* to do in respect of whales.

2 This Act authorises the US Secretary of the Interior to take measures to restore wildlife to parts of the US in which they have become extinct, and to regulate the introduction of wildlife (either American or foreign) into areas where they do not occur naturally.

Parker writes that by 1985 all ivory producing and major consuming nations were CITES parties, but there were still 'three entrepots that bucked the system'. These were Singapore, which was not a party 'and placed no restrictions on importing and exporting ivory'; Dubai, although 'nominally a party to the Convention'; and Burundi which:

> was altogether different. Where Singapore and Dubai were wealthy, Burundi was one of the world's poorest states, and surrender over ivory could have had serious general economic implications. Consequently Burundi resisted pressure to stop trading ivory and was averse to joining CITES.
>
> (Parker, 2004: 286–7)

This provides an interesting case study of how international pressure affects different states differently. It would appear to be as difficult to persuade a poor state as to persuade a rich one to act in a way that is not directly in its, at least short-term, interest; determining the reasons for state actions, and the factors that influence particular states in particular situations, is not easy.

Meredith tells us that the arguments 'raged anew' at a meeting of the 'Elephant Specialist Group' held in Kenya in 1987. A strong case for supporting the trade was made by Rowan Martin, a scientist from Zimbabwe who argued that 'high prices for ivory, far from threatening elephant populations, provided the means to protect them. What was needed in Africa, he said, was an effective system of protection and utilisation, like the one that Zimbabwe operated'. This remains the Southern African view in the twenty-first century. But, continues Meredith, 'CITES did nothing about the trade. The reason was that the prevailing policy within the world's conservation establishment – including the IUCN, the World Wildlife Fund and CITES – was to support sustainable utilisation' (Meredith, 2001: 214–6). In a sense, then, it can be argued that sustainable utilisation *was* the dominant idea, and that it went 'underground' for some years, but is now returning.

On the growing feeling that a ban might become necessary, and on increasing awareness of the full implications of poaching, I and O Douglas-Hamilton write that by the end of 1987, Western 'would not yet go for an ivory ban, wanting instead further dialogue with traders, but he threw his full weight into confirming that the trade was the principal cause of elephant decline', and they credit Western with coining a 'splendid phrase' referring to elephants as 'flagships for conservation' (1992: 306).

According to I and O Douglas-Hamilton, experience showed that 'any value attached to tusks would result in them being over-exploited' (1992: 316–7). They offer a rationale for this, and in so doing raise an interesting point about Japanese strategies in international law. 'The main economic force', they write, 'behind the ivory rush was revealed to be the increased prosperity of ordinary citizens of Japan, who could now afford to buy the ivory *hanko* seals, prized status symbols with which they signed their letters'. The dominant position of the Japanese in the trade, they say, 'gave them the power to make or break any attempts at an international ban, and it was widely believed by conservationists that the Japanese were unreceptive to global ecological concerns'. However, in this, they write, 'the conservationists were wrong'. They record that:

> a delegation from the Ivory Trade Review Group went to Japan to argue the elephant's case. The Japanese politely listened to the arguments, weighed the evidence and came to the conclusion that they should join the Western nations in the immediate ban pending the next CITES meeting in October, when the whole vexed question of elephants and the ivory trade was due to be debated. (ibid.: 325)

While the Japanese agreement to join the Western nations was to be short-lived, it is significant that it happened at all. The cynic might argue that it was merely a strategic move adopted in Japan's interest, but that it happened does suggest that Japan can be persuaded on occasion to act, apparently, in consort with the Western nations and against its own usual position. This might be a useful insight, given that Japan is often depicted as being obdurate.

The UK and the US continued to act 'unilaterally', under growing pressure from environmental NGOs with the US pushing a reworked African Elephant Conservation Act through Congress. This Act gave the CITES quota system a limited time to prove itself, after which 'a total US ban would become mandatory'. The Act banned ivory imports from nations (like Somalia) which had records of illegal ivory trading, and also 'threatened to ban ivory from all intermediaries'. Momentum was gathering for a ban. In May 1989, the UK 'took domestic action, followed by France and then the whole European Community. By mid-June, even Japan had decided to stop importing worked ivory' (I and O Douglas-Hamilton, 1992: 318, 330). Leakey moved closer toward direct involvement, Poole writing that she could not 'think of anyone else I knew who would risk his job and perhaps his life to attack the government openly on the ivory issue' (1996: 201).

Many different organisations and persons have taken, or been given, at least partial credit for proposing the ban. What was needed, however, was a state party to CITES to put the motion forward, and this ideally would be a range state. Many NGOs that might not otherwise have wished to call for a ban found that it was the only way in which to raise donor funds (Bonner, 1993). Politics and special/vested interests began to come to the fore. Arguably, what happened was that Western ideals were increasingly imposed, carried in on a tide of emotion. The World Wildlife Fund (WWF), for instance, is a hugely powerful organisation, but it appears to have found itself in a very difficult position. According to Bonner, it is 'committed to the principle of sustainable utilization, and it funds programs which encourage African communities to benefit through hunting and the sale of wild animal products', but it 'fears that most people donate money for conservation because they want to see animals preserved, not utilized' (1993: 62–3). It might even be argued – cynically – that the 'preservationist' view is not *really* the West's line at all, and that what the West really wants *is* sustainable use – as long as *it* controls the use, and decides *what* the use will be.

Lapointe, however, comments on WWF, scathingly, that 'while in appearance, WWF is supportive of sustainable use and supports positive conservation programs around the world; it does not recognize the universality of the principle of sustainable use'. For WWF, he says, 'the notion of sustainable use is fine for certain species, but certainly not for the "charismatic" species such as whales and elephants, which represent a major portion of its income'. Lapointe explains that in 1989, 'WWF completely revised its stance on the proposed ban on ivory trade, purely for financial reasons, and against the advice of its conservation department' (2003: 38). On the motives of NGOs, Bonner suggests that 'elephants had become the environmental fad' and he writes that the fact that:

> there was a schism as big as a canyon between the approach to conservation taken by the Africans on the one hand and the conservation organizations on the other was not surprising, not when one looked at the conservation organizations: they were the monopoly of white Westerners.
>
> (1993: 81)

Leakey writes that 'eliminating the market seemed such a logical thing to do that I was surprised to discover that the idea was highly controversial. Indeed, until 1988, very few conservation organizations had called for a ban on the ivory trade'. Leakey claims, in fact, that the idea 'wasn't popular with all conservationists or with African governments, which depended on the sale of ivory for income. In fact, until my appointment, Kenya had been against the ban' (Leakey and Morell, 2001: 85–7).

It is not always easy to work out exactly what certain protagonists stand for. Some may even not have definite views. After all, perhaps a definite view is not even realistic – or 'honest'. If one tries genuinely to understand the debate, then one might have to adopt a flexible position and admit exceptions, exemptions, local or regional idiosyncrasies, and that compromise is necessary. It might surprise some to learn that it was not, in the end, high-profile Kenya which proposed the ban on the ivory trade in 1989. Poole offers a reason, writing that:

> Kenya was, at that time, preparing to sell twelve tons of ivory, most of it confiscated from poachers, and it would have been impossible for the country to propose a ban on ivory when at the same time it was intending to sell a huge consignment.
>
> (1996: 222–4)

It was instead Tanzania, supported by Kenya, which eventually put the proposal forward.

Leakey had now been appointed as the new director of Kenya's Wildlife Service, and he decided that Kenya would not sell its stockpiled ivory. Other countries began to give support, including the US. In a bizarre twist, Poole writes, the head of Somalia's Wildlife Department even agreed, without the knowledge of his superiors in government who actively traded in ivory from Kenyan elephants, to propose the Appendix I listing (Poole, 1996). Little appears to have been straightforward, and we might never know precisely for what reasons important state and other actors acted as they did.

According to Bonner, Leakey's role in the movement was 'overstated', but he 'lobbied hard' and, 'if he had not, it might not have come to pass' (1993: 133). A point which needs to be understood *strongly* about international law is that, although it is about states, it is driven by individuals – which is, of course, why it is important to read and consider their accounts, no matter how self-serving or dismissive of others they might sometimes be. I and O Douglas-Hamilton write that the burning was 'Richard Leakey's inspired but sad finale to put an end to thousands of years of elephant killing and ivory collecting' (1992: 332). On Kenya's recent position, Western explains that Kenya has shifted its policies since 1989. He is critical of the burning of ivory, arguing that, today at least, it does not serve a useful purpose. He explained also that Kenya's 1989 burning may have occurred for reasons other than those traditionally put forward – in that there may have been promises from various pro-ban parties of massive compensation for destroying the ivory stockpile. The position was that there 'had been commitments made' – to approximately $50 million – by donors, if Kenya supported a moratorium; but this commitment was not fulfilled. In 1994/5, he says, Kenya burned ivory to show the donors without expecting a return; rather, not wanting the ivory to become an issue in the trade, the burning was intended to remove temptation and to draw attention to the fact that commitments had not been met. This burning was of about 12 tonnes, with the failing donors being the EU and the US, at governmental level (Western, pers. comm., 2004).

It is also worth mentioning at this point that, subsequent to the high-profile burning of ivory in Nairobi in 1989, there was a burning of Zambia's stockpile in 1992.[3] This was a publicity gesture that sank without trace, with virtually no apparent impact on conservation principles, and which implies that the Kenyan burning might have been a one-off publicity stunt – arguably the greatest deliberate public relations 'stunt' in environmental history. I have not met many people from conservationist or government circles in South Africa who do not mock the 1989 burning as having been 'burning money'.

Leakey explains that Kenya faced a number of opponents, particularly Botswana, South Africa and Zimbabwe, of which three countries he writes that he found them to have an 'almost visceral attachment' to their policies. He writes of a meeting of the CITES 'African Elephant Working Group' (a group formed in 1987) in early July of 1989 in Gabarone, where it was apparently 'claimed that Zimbabwe's elephants had an astonishingly high reproductive rate, the females giving birth at nine years of age' – Leakey's tone being firmly disbelieving (Leakey and Morell, 2001: 88–91). I think it is worth mentioning here because the same claim has been

3 It appears that Zambia burned its stockpile in 1992, on the strength of promises by NGO donors of compensation. Whether these commitments were met or not is in dispute (CITES, 2002). The point I am making here is that the incident is virtually unknown.

put forward by pro-whaling states in respect of minke whales. What is there to be said about the argument? I think that biologically it makes little sense. The question that might be asked is whether the fact that it is a common argument indicates similarities in the arguments in respect of both species, and therefore similarities in the views of the 'pro-use' protagonists?

According to Meredith, 'African opinion turned too. Tanzania, after losing 200,000 elephants in ten years, two-thirds of its national herd, became the first African country to support an Appendix I listing'. Opposed to this, Southern African countries 'fought tenaciously to prevent a ban', arguing that their herds were safe, they were expanding and that they represented a valuable economic resource of benefit to rural communities (Meredith, 2001: 221). Southern African countries today generally remain firmly opposed to the complete ban on the ivory trade. There is also a point here which should not be understated: that Kenya today no longer has very many elephants. While not for a moment downplaying the importance of preserving East Africa's elephants, there might be an argument to be made that it is the Southern African states which are now the only *true* African elephant range states.

As the CITES CoP approached, some support apparently grew for an 'accommodation' of the Southern African position, which would be achieved by way of a 'split listing': the elephant populations in Kenya, Tanzania and other countries where poaching was rife going onto Appendix I, and those in Botswana, South Africa and Zimbabwe remaining on Appendix II. The alternative was then suggested that the elephant would be declared an endangered species and listed on Appendix I, but that a country could have its population 'downgraded' to Appendix II 'if a panel of experts determined that the country was managing its elephant populations well and controlling the movement of ivory out of the country' (Bonner, 1993: 151). It was Somalia which formally introduced this proposal, and the eventual vote was 76 in favour and 11 against, with four abstentions – one of the abstentions being Japan (Bonner, 1993). That it was Somalia, the country which had done so much through its national poaching in Kenya's game parks to damage elephant populations in East Africa, that brought the CITES parties to their impasse and prompted Kenya to its act of burning, which proposed the compromise 'solution', is, of course, supremely ironic – but such is the nature of international law and politics!

Not all was rosy. Bonner points out that 'the most critical question still had not been answered: who would provide the money for elephant conservation?' (1993: 158). Thus Southern Africa found itself 'subsidising' the East – and indeed the rest of the world. This must raise important questions about the elephant as a symbol. The importance of the ban in this regard must be questioned, and also the questions of finance, of valuing the environment and of who should pay for the preservation of 'world heritage'. Developed states may not be willing to face the real costs of valuing the symbol.

Sifting through the rhetoric, there are some appalling examples of political self-interest to be found. American hunters, according to Bonner, 'feared, with good reason', that if the elephant were 'placed on CITES Appendix I, they would not be able to bring trophy tusks back into the US, from Zimbabwe or any other place'; and so hunting organisations in the US lobbied against an Appendix I listing. Then, he says, these organisations 'cut a deal with groups that wanted a ban on ivory trading: the hunters would not work against the ban, in Washington or with other countries, if the pro-ban people would do nothing to jeopardize sport hunting of elephants' (Bonner, 1993: 268–9). This will not be the last occasion where US politics, vested interests and actions in its own interest, coincide – hypocritically. The major Western NGOs do not always come out from under scrutiny favourably either.

Kreuter and Simmons write that 'politically, the voting patterns of the 91 member nations attending the 1989 CoP are revealing'. They explain that 'of the 76 nations endorsing a trade ban only 20% (19) had elephants while 63% (57) were non-African countries' (Kreuter and Simmons, 1994: 45–6); whereas, in contrast, 73 per cent (8) of the 11 countries opposed to the ban had elephants. These statistics, the authors suggest, show that it was countries with no elephants or with rapidly decreasing elephant populations which voted to 'impose an anti-trade policy on countries with conservation programs under which elephants populations were stable or increasing' (ibid.). Gavron writes of the 1989 CoP, however, that 'in the end, there was little disagreement' and 'even Japan voted in favour' (Gavron, 1993: 144). In fact, Japan abstained rather than voting in favour, but the effect was the same.

It is significant to note that both the UK and the US received exemptions that suited them. The US won the sport trophy exemption it needed to keep its domestic hunting lobby happy, and the UK won a 6-month exemption for Hong Kong to dispose of its ivory stockpile (I and O Douglas-Hamilton, 1992). It would be no wonder if range states were to look askance at these two developed countries and question their altruism in the debate over 'saving' the elephant.

5.2 The next CoPs

Leakey describes the Southern African countries as telling the press that they were 'going to do everything possible to overturn the decision at the next conference in three years, and that they might continue to trade outside the ban with those countries not signatory to the CITES convention'. They continued to argue, he writes, 'that they felt that they were being unfairly punished for successfully managing their elephants' (Leakey and Morell, 2001: 119). This is the argument that Southern Africa subsidises East Africa. However, nothing in the ivory debate is straightforward and the management by the Southern African countries might not have been as successful as they clearly believed it to be. It seems that there is always a potential for poaching, just below the surface, a potential that can be set off even by half-signals. Generally, in the years immediately following the 1989 CITES CoP, it appeared that the ivory trade had collapsed – there were even reports of further burnings of ivory stockpiles, none of which, of course, had anything like the impact of the 1989 Nairobi burning. I and O Douglas-Hamilton wrote in the early nineties that 'Taiwan, once feared as a potential ivory loop-hole country, has twice burnt its ivory' (1992: 345). There have been several mentions by commentators of burnings subsequent to the 1989 Nairobi burning – vide Zambia's 1992 burning, and this mention of Taiwan. Details for all of these are, however, elusive, and the Nairobi burning remains the only one that was of major significance.

Hanks writes that 'in South Africa the gross income from the sale of elephant products over the five year period from 1985 to 1989 averaged US$1.4-million per year', and that 'these funds accrued directly to the National Parks Board, and helped to offset the costs of security and management of the various reserves under their control, notably Kruger National Park'. In June 1991, he records, Botswana, Malawi, Namibia, Zambia and Zimbabwe established the Southern African Centre for Ivory Marketing (SACIM) (Hanks, 1997: 23).

CITES would meet again in Kyoto in March 1992, and Zimbabwe, per Leakey, 'worked hard to convince the delegates to remove the elephants from Appendix I so that the ivory trade could resume'; but, he suggests, 'our security data showed that the ivory traders were simply waiting in the wings for the illegal trade to restart. If the Zimbabweans had their way, Kenya would once again have a poaching war on its hands' (Leakey and Morell, 2001: 218–9). This seems still to be roughly the position today and, in effect, both sides accuse the other of selfishness. According to the Southern African countries, they lose revenue by being forced to compensate for others' inadequacies. According to the East African range states, the Southern African countries are determined to profit, even at the knock-on cost to other states.

Bonner writes that there was 'no serious movement at Kyoto', 1992, to downlist the African elephant generally, but that the Southern African nations wanted to have their elephant populations downlisted to Appendix II, which meant they would be allowed to trade in ivory. However, he explains that, as they knew the strength of the opposition to ivory trading, the Southern Africans 'were willing to agree to a moratorium for two years' (Bonner, 1993: 281).

At CoP 9, in 1994, according to Hanks, South Africa submitted a proposal to downlist its elephants to trade internationally in non-ivory products – 'mainly skins from elephants culled for ecological reasons' in the Kruger National Park. A commitment was made that funds raised would go back to park management. Hanks explains that 'in the five years leading up to the Appendix I listing coming into place in 1990, the Park had earned US$340,000 annually from the sale of skins'. An integral part of South Africa's proposal, he explains, was the commitment that, if the proposal was accepted, the country would withdraw its reservation to the Appendix I listing. The implication of this being that South Africa 'would never again be able to trade in ivory internationally without the approval of CITES'. Those opposed to the South African proposal, however, expressed concern that placing South African elephants on Appendix II, 'even for non-ivory products, would act as a signal for the re-opening of the ivory trade, and would thus stimulate further poaching, although no evidence was brought forward to substantiate this'. Ultimately, though, opposition to the proposal, records Hanks, 'continued in the build-up to the CoP, leaving South Africa with little alternative but to withdraw the proposal at the CoP itself' (Hanks, 1997: 23).

5.3 CITES and the African elephant

Some efforts have been made roughly to quantify the impact of the closure on the Southern African range states. Bonner writes that some calculations showed that 'by depriving countries of ivory sales, the ban was the equivalent of a $50-million tax on African governments' (1993: 142). While it will never be possible to quantify such impact accurately, there being 'few mechanisms to retrieve or capture non-use values' (Blignaut *et al.*, 2008: 468), this might give some idea of its magnitude.

Despite their initial threats to trade ivory outside of CITES, Zimbabwe and South Africa did not (Chadwick, 1992: 344). Zimbabwe and South Africa clearly decided not to risk the adverse consequences of being seen as ignoring international law, although as clearly they resolved privately to regroup and fight again.

Kreuter and Simmons write that the 1989 vote in favour of the Appendix I listing 'subverted' the CITES treaty as the voting parties were aware that several elephant populations were not endangered. Their conclusion is that the majority of the voting parties therefore 'wilfully violated the CITES guidelines, thereby denying southern African states (with stable or increasing elephant populations) the protection of the legal instrument they had undertaken to uphold'. This rhetoric is similar to the charge that has been levelled against the anti-whaling parties within the ICRW. Kreuter and Simmons argue further, on the basis of illegal trade having occurred in 1993 and 1994, that the listing of the elephant on Appendix I may not have been altogether successful and that 'internationally the demand for ivory has not been eliminated' (Kreuter and Simmons, 1994: 46–7, 47). The implication being, then, that if not successful, the 'harm' which the ban does to the Southern African interests is not outweighed by any benefit.

It appears, however, that only two other nations among CITES's then 112 members, Japan and Switzerland, supported downlisting, and the Southern African bloc withdrew its proposal. 'But', per Chadwick,

> they did not agree to abide by the will of the majority. Instead they left threatening to establish trade in elephant products – including ivory – among themselves and the more than fifty nations that are not signatories to CITES and do not abide by international regulations on commerce in wildlife.
>
> (1992: 466–7)

Interestingly, though, at the time of writing only one country retains a reservation in respect of the CITES Appendix I listing of elephants – this being Malawi.

In June of 1997, I was in Harare at the time of the 10th CITES CoP. It was a fascinating place to be, to observe the hysteria and hype that surrounded the elephant issue. Approximately 75 proposals to uplist or downlist various species were put forward; however, the main issue in the public eye at the CoP was that of the African elephant and its possible downlisting. The line taken by Southern African countries was fervently in favour of downlisting. Zimbabwean President Robert Mugabe suggested that elephants, 'especially because of their huge bodies, consume large amounts of underground water and, we believe, every species must pay its way to survival' (CITES, 1997). Although somewhat emotional, comments such as this may indicate deep resentments. Zimbabwe certainly felt hard done by and Botswana, Namibia and South Africa took the same line.

5.4 CITES, the African elephant and whales

Japan and Norway supported the Southern African countries, arguing that countries have an inalienable right to the sustainable utilisation of their natural resources. Few other developed countries gave such support, and it is worth considering that these two had an agenda other than altruistic concern for the Southern African viewpoint. There is obviously more to be said for a country's right to profit from natural resources found within its borders than for a country's right to utilise resources from the high seas, or 'global commons', against the wishes of other states. However, the principle of sustainable development remains the same and, for the pro-whaling states, downlisting of the elephant would provide a useful precedent.

I know only a few other writers who have drawn the link between elephant and whaling issues – either at CoP 10 or generally. Payne wrote in 1998 that:

> the circumstances of the vote, taken in the last evening of the meeting at the insistence of Japan, Norway, and Zimbabwe, were unusual. For the first time ever in CITES, the voting was secret. The secrecy was in itself a controversial decision, also pressed by the delegates from Japan and Norway.
>
> (1999/1998: 270–1)

While an interesting comment, it is not true that this was the first ever secret ballot – the first secret ballot was held at CoP 4 in 1983. What is very interesting, however, is that the *vast* majority of secret ballots held at CITES have concerned either elephants or whales. In the plenary sessions and two committees, from CoP 9 to CoP 12 (i.e. 1994, 1997, 2000 and 2002), all but one of the secret ballots held related to elephants, mahogany or marine animals (CITES, 2004).

The result of the ballot, writes Payne,

> surprised many participants, suggesting that under the cover of secrecy some representatives had cast their votes contrary to their openly stated positions and to their mandates from their home constituencies. This apparently allowed several countries with no elephants and no trade in or industry using ivory – Norway, for instance, and a group of Caribbean states

to give support to the ivory trade. This would indeed be an interesting voting bloc – and a clear example of the linkage of the ICRW with CITES, and more specifically of the elephant with the whale. 'In what seemed to be a reciprocal action', continues Payne, 'several Southern African nations with no trade in or industry using whale products then supported Japan in trying (again by secret vote) to reverse the protected status of the North-East Atlantic population of minke whales' (Payne, 1999/1989: 270–1).

Darby has suggested that Japan and Norway 'restrained themselves' for years from using their lodged objections and 'taking a fight from the IWC into another organisation'. He suggests also, however, that the two countries have tried 'repeatedly' to have minke stocks downlisted for trade, which seems on the contradictory side (Darby, 2007: 187). Friedheim, writing in the context of whaling, suggests that 'as important a test as whaling is of the viability of multilaterally established rules, what is happening within the IWC is only one example of a more general revolt against dominant powers imposing their will on others'. He then refers to CITES CoP 10 and the partial downlisting of elephants, and draws the conclusion that:

> perhaps the partial accommodation of the needs of these states is a sign that those in control of international resource management issues have become more sophisticated in their approach and can take care of local needs and sensibilities, and this is a sign that a serious threat to global environmental governance can be avoided.
>
> (Friedheim, 2001b: 19–20)

Of other commentators, Vogler does suggest that at CoP 10 the:

> conflict within the IWC spilled over into CITES as Norway and Japan sought to weaken the whaling moratorium by proposing the 'downlisting' of certain whales from Appendix I. If the move had been successful it would not only have legalized trade in whale products but would have had clearly detrimental implications for the authority of the IWC.
>
> (2000: 52–3)

Pickover explains that in 1997, Japan and Zimbabwe, 'in particular, mounted a very successful propaganda campaign' to argue that the ban was essentially an 'anti-African plot from Western elitists who did not understand the need for ivory to bring in cash to pay for conservation and social services'; and concludes that 'there also appears to have been a trade-off between the Norwegians and certain African countries' in the form of 'we will let you kill elephants if you will let us kill whales' (2005: 74–6).

How deliberate a strategy this may have been, and how deliberate the cooperation, will probably never be known. In the run-up to the Harare CoP, however, Greenpeace certainly appeared to believe that a conspiracy was afoot[4], adding that the 'other important downlisting proposal to be discussed in Harare

4 The NGO alleged in 1997 that six countries (Botswana, Canada, Japan, Namibia, Norway and Zimbabwe) were leading the 'assault on CITES' through an alliance to 'change CITES from a conservation convention into a wildlife trade one'. The claim was then made that 'in January the German weekly *Der Spiegel* published leaked material showing that a government representative from Zimbabwe was working secretly with other ivory-producing states in Southern Africa to weaken CITES' (Greenpeace, 1997).

is on whales' and that the 'wording of the documents, submitted by Japan and Norway, leaves little doubt that theirs is a concerted effort, which is part of a wider campaign for the reopening of whaling'. This, the NGO concluded, 'could allow the whalers to copy the would-be ivory exporters: at a minimum downlisting with zero quotas now, followed by attempts at increasing these at the next CITES conference' (Greenpeace, 1997).

In Harare in 1997, South Africa proposed a compromise arrangement, which was initially turned down by the CITES CoP. However, out of this refusal came the formation of an inter-governmental working group which put forward amended proposals. Put to a secret ballot, these proposals were adopted with the necessary two-thirds majority. The elephant was not downlisted worldwide, but a 'once-off' test auction of ivory was approved. The sale was to be in the total amount of 59.1 tonnes (of a combined stockpile of more than 150 tonnes) from the three countries; to be sold only to Japan; to be used only for the purpose of the manufacture of *hanko*; and to take place only after a 21-month moratorium, and after an assessment by a monitoring/investigating committee that was to be set up. This committee was to be called MIKE (Monitoring Illegal Killing of Elephants).

Many of the restrictions which were put in place, and conditions which had to be met, were glossed over in media reports, however, and the word appears to have gone out immediately that restrictions had been lifted. A few months later, while working as a field guide in a safari lodge in Zambia, I was told that poaching of elephants had increased in that country almost immediately after the CITES CoP had ended. Meredith comments that 'the role of the ivory trade remains unresolved' and the pressure is becoming greater all the time. According to Meredith, annually most countries in Eastern and Southern Africa add at least 1, and in some cases 5, tonnes to their stockpiles; Zimbabwe accumulating about 10 tonnes a year (Meredith, 2001: 224–5).

In 1999, the proposed auction of ivory (by Botswana, Namibia and Zimbabwe) did in fact go ahead, despite no proper report having been put forward by MIKE and despite evidence that poaching had in fact increased. It seems, though, that the auction itself was successful. An amount of approximately US$5 million was apparently received from the auction, a substantial amount of which was earmarked for elephant conservation (CITES, 1999). The charge has been made, however, that the conditions laid down 'were badly drafted and ambiguous, and had been agreed in questionable circumstances'. According to Reeve, in February 1999 all countries were approved, except Botswana with respect to its ivory recording system. It had been one of the requirements for the auction sale to go ahead that the states involved join the Lusaka Agreement.[5] Reeve records, however, that:

> none of the three range states had joined the Lusaka Agreement. Instead, they offered as evidence of compliance their involvement in the development of a

5 Lusaka Agreement on Cooperative Enforcement Operations Directed at Illegal Trade in Wild Fauna and Flora, 1992.

> draft protocol on cooperation in wildlife management under the SADC, together with their engagement in bilateral agreements establishing joint commissions with neighbouring countries on defence and security.[6]

The Secretariat, says Reeve, however, 'verified this as adequate to comply with the condition'. Not all agreed with this decision. Eight African range states apparently 'raised serious concerns about compliance with the conditions in a letter to the Standing Committee' (Reeve, 2002: 77–80).

One 'clarification' made was that 'implementation of MIKE was not necessary for the ivory sales to go ahead'. Instead, a temporary mechanism was agreed to, which would allow trade to be stopped if it should happen that information from national reporting indicated an 'important increase' in illegal hunting or trade, (however, the phrase 'important increase' was not defined). Another 'clarification' made was that the Lusaka Agreement was considered to be not the only mechanism, but just 'an example' of such a mechanism (Reeve, 2002: 77–80).

5.5 Subsequent CoPs

At CoP 11, in Nairobi in 2000, a decision was taken on downlisting of the elephant. The South African elephant population was downlisted from Appendix I to Appendix II, on the condition that no international trade would take place until after CoP 12 in 2002 – the wording being that a 'zero export quota for raw ivory' was put in place.

There must be virtually no market in Norway for ivory, and ultimately the trade in ivory cannot be seen as being in itself of crucial economic importance in Japan's eyes either (Miyaoka, 2004, records a steady decline of Japanese demand). Japan has endured the loss of supplies of ivory since 1989. The quantities of ivory which Japan can import legally under CITES supervised arrangements are arguably not of huge significance. In many ways, however, the ivory trade is symbolic of conflicting interests that are the subject of much controversy at CITES CoPs. The elephant controversy of 1989 was 'merely a harbinger', suggests Ong, for 'a wider and even more acrimonious debate over the appropriate environmental conservation paradigm to be employed by the CITES regime for the future – sustainable use or preservation of species'. Within this wider debate, Ong argues, the Southern African elephant range states:

> have emerged as the leading proponents of the sustainable use paradigm, which emphasises the goal of using living resources to meet both human and ecological needs, and asserts that in many cases wildlife can best be conserved by exploiting it for economic gain.
>
> (Ong, 1998: 300)

6 This draft protocol was eventually to become the SADC Protocol on Wildlife Conservation and Law Enforcement, 1999.

It might not be a good thing that so much attention is given to the elephant – insofar as it distracts from other species. At the same time, this illustrates the crucial importance to conservation globally of the elephant. If the elephant is a battleground for conservation, and more resources are to be sunk into its conservation than into the conservation of other species, then it is of huge importance that the optimal result be achieved. In this way, 'lesser' species might ultimately benefit.

CITES CoP 12, 2002, was held in Santiago. Southern African countries again sought downlisting. Zimbabwe did not receive permission for another auction sale, but Botswana, Namibia and South Africa did. South Africa gained the right to sell most (nearly 28 tonnes) of its stockpile of 32 tonnes. The decision provided for a 'one-off purchase for non-commercial purposes of government stocks declared by African elephant range states to the CITES Secretariat within the 90-day period before the transfer to Appendix II of certain populations of the African elephant takes place' (CITES, 2002).

The sale was made subject to conditions such as that only whole tusks (marked and registered) would be exported; confiscated ivory (or ivory of unknown origin) would not be exported; only countries meeting certain conditions (set by CITES) would be eligible to purchase ivory; the export of ivory would take place through a single government-controlled centre; and that all net revenues from the sale of ivory would be used for projects that promote the conservation of elephants. The eventual approval of the proposal included the stricture that the ivory stocks declared should be marked in accordance with the ivory marking system approved at CoP 10 – and that an independent audit of declared stocks should be undertaken under the auspices of TRAFFIC. There was opposition, such as from Germany, which indicated that it 'would review funding support for MIKE if the elephant proposals were accepted' (*IISD*, 2002). Botswana's proposal was, however, accepted with 59 in favour, 26 against and 21 abstentions; Namibia's proposal passed with 65 in favour, 28 against and 22 abstentions; and South Africa's proposal passed with 65 in favour, 24 against, and ten abstentions. All were secret ballots.

Interestingly, Tanzania spoke openly in support of South Africa's proposal. For Tanzania to break ranks at CITES with countries such as Kenya and India on the African elephant is a clear indication that the trend worldwide is toward sustainable use of natural resources – and that the ivory trade, albeit heavily restricted, will probably re-open on an ongoing basis within years soon to come. A move of potentially huge significance was to occur in 2003, when it was reported that Tanzania had indicated that it would like to sell off some of its 80-tonne ivory stockpile (Mkoka, 2003). Tanzania would even in 2007, 2010 and 2013 make such proposals formally to the CITES CoP – before withdrawing them.

Despite receiving some support, including interestingly from Antigua and Barbuda, Zimbabwe's proposal was, when voted on in a secret ballot, rejected with 60 in favour, 45 against and ten abstentions. When it introduced its own proposal, Zambia explained that it lacked financial support and that it needed ivory sales to raise revenue. The proposal was rejected with 57 votes in favour, 54 against and seven abstentions. Stating that they did not wish 'to target Zimbabwe', India and

Kenya withdrew their joint proposal on uplisting the African elephants to Appendix I (CITES, 2002).

Some recognition of the need for funding incentives to be provided to the developing world in international environmental treaties can be seen in Resolution 12.36, CITES CoP 12: 'parties, donors and organisations are requested to provide urgent financial and technical support to strengthen the implementation of Resolution Conf. 10.10 (Rev. COP 12) regarding control of the internal ivory trade in elephant range states'. Kenya, however, opposed proposals at CoP 12 to re-open the ivory trade and warned that it will continue to press for increased restriction of the ivory trade. Clearly, consensus on the re-opening of the ivory trade would continue to be controversial.

Linkage between CITES and the IWC was raised in the US in April 2004, when the Fish and Wildlife Service addressed a House Resources Subcommittee on the forthcoming CITES CoP. Its representative said that the Service anticipated that there would be 'several proposals dealing with whales'; explaining that 'this is a very contentious issue that has a long history within CITES'. Proposals, it was suggested, 'to downlist stocks of both Bryde's whales and minke whales from Appendix I to II by Japan have been defeated at the last four CITES meetings', and it was then argued that 'if adopted, these proposals would re-open international commercial trade in whale products, and could foster increased poaching of protected whale species'. The US, the representative said, 'continues to be strongly opposed to the downlisting of whale species in accordance with the commercial whaling moratorium of the IWC', and it was then explained that the US 'believes' that CITES should honour the 'request for assistance in enforcing the moratorium which was communicated by the IWC to CITES in 1978'. In conclusion, something of a sop to supporters of sustainable use was given with the advice that the US is continuing 'to participate in the IWC efforts to develop a Revised Management Scheme (RMS) that includes an effective inspection and observation scheme for use in the event that the moratorium on commercial whaling is lifted' (Stansell, 2004). Supporters of sustainable use would probably argue, however, that the US has no genuine intention of completing the RMS at any point in the near future.

Despite the confidence of the Southern African states, it was reported in September that there might not be as many elephants in Southern Africa as previously thought. In this regard, it was alleged that Zimbabwe might have deliberately inflated the figures for the country's elephant population. The media report suggested that the real motive behind such inflation might be not so much resumed ivory trading, as politically well-connected persons looking for 'an opportunity to cash in on the ivory stockpiles they have accumulated from the indiscriminate shooting of animals in seized private game parks and in national parks' (Peta, 2004).

In October at CoP 13, CITES rejected a Namibian proposal for an ivory export quota to allow it to export 2 tonnes of ivory annually from elephants that died of natural causes, but did approve the country's 'commercial trade' in elephant hair and skin. The difference here was that, previously, Namibia had been permitted to trade in elephant hides for non-commercial purposes only. Likewise, a proposal by South Africa to allow trade in elephant skin was approved. Namibia, however, was 'denied a request to sell carved elephant ivory encased in jewellery' (*News24.com*, 2004). Kenya, along with several other Central and West African countries, had opposed the Namibian proposal, and had unsuccessfully proposed having one-time ivory stockpile sales postponed for a period of six years.

In March 2005, Botswana's President Festus Mogae, according to 'a foreign ministry official', was 'expected to ask Zambian President Levy Mwanawasa for support to downgrade the status of elephants under CITES'. The official apparently said that Botswana has an overpopulation of elephants and that Mogae 'wanted to move elephants to a different category under CITES that would allow them to be culled' to 'acceptable levels' (*News24.com*, 2005). This appears to be a fairly common misunderstanding – that CITES prevents culling. CITES does not. A country may, if it desires to, cull as many elephants as it likes. That would simply be a domestic management decision. The underlying logic, or rationale, appears to be, however, that it is simply impossible to cull if one cannot then sell the ivory.

In October 2006, it was reported that the Standing Committee (SC) of CITES had rejected a request by Botswana, Namibia and South Africa to proceed with the one-off sale of stockpiled ivory that had been provisionally approved in 2002. The matter would next be considered at the next meeting of the SC – the function of which is to oversee the implementation of CITES decisions between CoPs – in May of 2007. According to CITES, the SC had decided that one of the conditions, in terms of which exports of ivory by Botswana (20 tonnes), Namibia (10 tonnes) and South Africa (30 tonnes) had been allowed in principle, had not been met. According to a CITES press release, certain 'baseline data' had not been supplied (CITES, 2006).

Also arising from the SC meeting, it was reported that Finland, on behalf of the European Union as it held that body's rotating presidency, 'welcomed' the new Strategic Plan for 2008–2013 adopted by CITES, and noted that 'reducing the rate of global biodiversity loss should be the overarching objective of the Plan'. Japan, however, 'opposed the proposed Strategic Plan objective, which states that the SC has the primary role in promoting compliance and addressing non-compliance with CITES' (*IISD*, 2006). Clearly, reducing global biodiversity loss can best be argued as being within CITES's mandate if one sees the treaty as having changed its emphasis since its inception in 1973. At inception, the treaty might arguably have had this as a goal – despite condoning (if not actively promoting) trade and targeting certain species only – but it requires something of a stretch of the imagination.

5.6 Good CoP, bad CoP, compromise CoP

CITES CoP 14 was held in June 2007 in The Hague. In the lead up, Botswana and Namibia proposed that in future the elephant populations of Botswana, Namibia, South Africa and Zimbabwe be subject to a maintenance regime involving annual export quotas for trade in raw ivory; that this trade be restricted to trading partners certified by the Secretariat to have 'sufficient national legislation and domestic trade controls to ensure that the imported ivory will not be re-exported and will be managed as determined by CITES'; and that the proceeds of the trade in raw ivory be used 'exclusively for elephant conservation and community development programmes' (CITES, 2007). Botswana proposed, also, that in respect of its own population, trade in hunting trophies be allowed; also trade in hides, and in leather, for commercial purposes; trade in live animals for commercial purposes (to appropriate and acceptable destinations); trade annually in stocks of raw ivory (whole tusks and pieces, of not more than 8 tonnes) of stocks owned by Botswana's government and from Botswana, for commercial purposes, with trading partners certified by the Secretariat to have sufficient controls to ensure that the ivory will not be re-exported; and that there be a one-off sale of the Botswana government's stockpile (of not more than 40 tonnes), immediately after the adoption of the proposal, with trading partners certified by the Secretariat to have sufficient controls to ensure that the ivory would not be re-exported (CITES, 2007).

Another significant development in 2007 was the proposal put forward by Tanzania for transfer of its elephant population from Appendix I to II for the purpose of allowing: first, 'trade in registered stocks of raw ivory in whole tusks and pieces'; second, 'trade in live specimens for non-commercial purposes to appropriate and acceptable destinations'; and third, 'trade in hunting trophies for non-commercial purposes' (CITES, 2007).

Tanzania's proposal was a lengthy document. The rationale given in the proposal was that it aimed at 'promoting sustainable conservation of the elephant population in Tanzania' with the revenue generated being 'reinvested into wildlife conservation in protected and non-protected areas and supporting development activities of communities living within the elephant ecosystems' (CITES, 2007). Proposed restrictions included that the ivory traded would be registered and government-owned ivory originating in Tanzania only; it would be sold only to CITES-approved trading partners; the sale would occur only after the prospective importing countries had been verified by the CITES Secretariat, and the MIKE Programme had reported to the Secretariat on baseline information regarding elephant numbers and illegal killing incidents; and that the maximum amount sold would be 100 tonnes. The proposal then suggested that the Tanzanian elephant population does not qualify for Appendix I listing, per the relevant biological criteria; the population being not small, not restricted to an area and having increased (apparently from 55,000 in 1989 to 141,000 in 2006).[7] Further, according to the proposal, the Tanzanian government has put effective elephant conservation measures in place, including legislation, special training of wildlife officers, an elephant management plan, expansion of its protected areas network, provision for portion of revenues from sport hunting to go to community development projects, ratification and implementation of various international and regional agreements (including the

7 That the population is claimed to have nearly tripled in 17 years might raise a few eyebrows.

Lusaka Agreement, ratified in 1996, and the SADC Protocol, ratified in 2000). According to the proposal, culling was not envisaged in order to generate ivory for trade – despite the increasing elephant population – as the proposal was to allow the 'export of ivory stock in the custody of the Government of Tanzania'. The main rationale put forward appears to be that there has been increased human–elephant conflict since the late 1990s, and that putting ivory revenue back into communities would 'reduce the growing negative attitude by the rural communities towards elephants' (CITES, 2007).

The Tanzanian proposal was, however, withdrawn before CITES met. According to Jennifer Lonsdale, of the NGO Environmental Investigation Agency, the proposal was withdrawn because EIA convinced the Tanzanian government that poaching is ongoing (Lonsdale, pers. comm., 2007). While this might have been the case, Tanzania went on to put forward similar proposals in 2010 and 2013. Tanzania's moves are significant. That Tanzania considers its elephant numbers to be sufficiently stable, poaching to be under control, markets to be receptive and – perhaps most significantly – the general *zeitgeist* to have changed enough to have made the proposal is, in my opinion, an important indicator of the direction in which international regulation of the wildlife trade is heading.

On the other hand, Kenya and Mali submitted an annotation proposal to amend the populations of Botswana, Namibia and South Africa so that 'no trade in raw or worked ivory shall be permitted for a period of twenty years' – except for 'raw ivory exported as hunting trophies for non-commercial purposes', and for 'ivory exported pursuant to the conditional sale of registered government-owned ivory stocks agreed at' CoP 12. Further, the proposal was to do away with the permissible trade by Namibia of *ekipas* (traditional indigenous ivory ornaments) (CITES, 2007).

The various participants in the fraught negotiations toward agreement on ivory spent the first few days of the CoP locked in disagreement. However, a glimmer of hope for a deal began to emerge based on stockpile sales by the four Southern African range states, and a moratorium of approximately a decade. At this stage, though, none of the existing proposals were withdrawn by the various protagonists (*IISD*, 2007). The parties then discussed various matters, including enforcement and monitoring. Evidence was presented which showed that there was a decline in trade between 1999 and 2004, but then a sharp increase since 2005.

South Africa put forward an amended proposal, which combined the original proposals from Namibia and Botswana, Botswana, and Kenya and Mali. As amended, the proposal was to the effect that trade in hunting trophies, hide and hair be permitted, and that trade in registered government-owned raw ivory be permitted, subject to verification of trading partners, and with the proceeds being used exclusively for conservation and community development. Further, the proposal was that a one-off sale be allowed of a maximum of 70 tonnes of ivory for Botswana, 50 for Namibia, 40 for South Africa and 15 for Zimbabwe, in addition to the quantities agreed to at CoP 12; and finally that it be provided that no further trade take place before CoP 16, with the SC being mandated to take a further decision on export quotas and continuity of trade.

Namibia then called on parties to support the proposal, highlighting the livelihood and conservation benefits of trade in ivory, stressing that Southern African elephant populations 'are robust and should be treated under CITES provisions for trade in Appendix II species'. The European Union then 'reiterated' that, while elephants have 'global charismatic appeal' and other countries 'may help by facilitating consensus', the 'core issues should be resolved by African countries'. The EU delegate then proposed that there be an adjournment for further negotiations. This 'Africa should resolve' stance appears to have been a presaging of agreement, with the European Union perhaps preparing its citizenry for a backing away from European prescription of a preservationist view. In the run-up to the CoP, the European Parliament had 'called on' the Commission and EU member states to support the proposal by Kenya and Mali for a 20-year moratorium on all ivory trade, and had 'urged' the same bodies to reject the Southern African proposals to establish annual export quotas (European Parliament, 2007).

Kenya stated that 'finding common ground would be preferable to applying different solutions in different parts of Africa', and discussions continued with an informal ministerial consultation facilitated by Zimbabwe's Environment Minister (*IISD*, 2007). Chad and Zambia, for a number of African countries, then presented a compromise proposal to the effect of a one-off sale of raw ivory from government stocks registered before 31 January 2007 from Botswana, Namibia, South Africa and Zimbabwe, in addition to the quantities agreed to at CoP 12 and subject to verification of trading partners. Further, the proposal was that no further ivory trade proposals be submitted to the CoP for a period of nine years after the one-off sale, and that 'the SC may decide to stop trade in case of non-compliance or proven detrimental impacts on other elephant populations'. Further, the proposal included that the SC would:

> propose a decision-making mechanism for ivory trade by CoP 16, and review the status of elephants; that range states would develop an African elephant action plan; and that the Secretariat would establish an African elephant fund administered by the SC.

The proposal was eventually adopted by consensus – 'and acclamation' – and Botswana, the European Union, Kenya and South Africa withdrew their respective proposals (*IISD*, 2007).

After the CoP, the European Parliament (Committee on Environment, Public Health and Food Safety) explained that the consensus agreement, while authorising the sale as agreed, included a provision that the Secretariat may decide to halt trade 'in case of non-compliance or proven detrimental impacts on other elephant populations'. The agreement further included the development by the relevant African countries of an African Elephant Action Plan; this Plan to include prohibition of unregulated domestic ivory sales, enforcement of both new and existing legislation, promotion of public awareness, reporting of ivory seizures and any changes to legislation or enforcement measures and the creation of a new African Elephant Fund to support all of these measures. The Committee then described this consensus between African Ministers as 'historic', as ending '18 years of controversial debates in CITES' and as 'paving the way for more constructive dialogue on elephant conservation programmes in the coming years' – with discussion on a mechanism for decision-making on future sales to occur only after CoP 16 (European Parliament, 2007). In the result then, at CoP 14 a compromise solution was reached, in terms of which certain countries would be able to sell their stockpiles of ivory, but that there would then be a nine-year moratorium on sales. This was generally expected to have the effect of removing ivory from the agenda for some time. The expectation was to prove erroneous, however.

5.7 Recent CoPs

In July 2008, China and Japan were granted approval by the SC to be buyers of ivory from Botswana, Namibia, South Africa and Zimbabwe. While EU countries voted in favour, it appears that there was still significant opposition from some African countries, led by Mali (the vote ran nine in favour, three against, with two abstentions).

In late 2008, four auctions, one each in Botswana, Namibia, South Africa and Zimbabwe, were conducted under the supervision of the CITES Secretariat. Altogether, 102 tonnes of ivory were sold to accredited Chinese and Japanese traders – the total amount of money raised from the sales was US$15,400,000. The average price paid was US$157 per kg – apparently substantially below the alleged average price for ivory illegally traded in the preceding year (US$750–850). It seems that Namibia sold 7.2 tonnes for US$1,186,260, Botswana sold 43.15 tonnes for US$7,093,550, Zimbabwe sold 3.7 tonnes for US$500,000 and South Africa sold 47.35 tonnes for US$6,703,000 (CITES, 2008). According to the terms under which CITES authorised the auctions, the proceeds from the sales were to be used exclusively for elephant conservation and community development programmes 'within or adjacent to' elephant ranges. At the time of writing, it is still far too early (by several years, at least) to comment on the effectiveness of this sort of commitment. The cynical observer might comment that there are many ways in which such stipulations might be circumvented – such as the reducing of funds allocated from other sources to communities which benefit from funds raised by ivory sales.

After CoP 14 it was my understanding, and I certainly thought that it was the general understanding, that no further proposals on the ivory trade would be put forward for a moratorium period of nine years. However, in the run-up to CoP 15 in 2010, three proposals were put forward. Tanzania put forward a proposal for a downlisting in order to allow it a one-off sale of just less than 90 tonnes of stockpiled ivory, and Zambia put forward a proposal which would allow it trade in ivory as hunting trophies, trade in live animals, trade in hides and a one-off sale of almost 22 tonnes of stockpiled ivory. In opposition, a proposal was put forward by Congo, Ghana, Kenya, Liberia, Mali, Rwanda and Sierra Leone that no downlisting proposals concerning African elephant ivory should be made for a period from CoP

14 ending 20 years after the date of the sale of ivory in November 2008 – this to replace the nine-year period agreed to in 2007.

In Committee I, Tanzania put forward its proposal and noted that the nine-year moratorium decision taken at CoP 14 had applied only to countries which had elephants listed on Appendix II at that time. Zambia also put forward its proposal and, with Uganda supporting, asked the Secretariat to clarify the outcome of CoP 14, as there were differing opinions on the matter. The Secretariat replied that 'the record of that meeting was quite clear and that the moratorium applied only to parties whose elephant populations were listed in Appendix II at that time' (CITES, 2010).

Tanzania withdrew the ivory trading segment of its proposal,[8] leaving trade in ivory from hunting trophies, trade in live animals and trade in hides. The amended proposal was supported by Botswana, Senegal, South Africa and Uganda, and opposed by the Gambia, Ghana, Kenya and Rwanda. A vote was held, by secret ballot at Tanzania's request, and the proposal was rejected with 55 votes for, 55 against and 34 abstentions. Zambia then asked that debate on its proposal be opened, 'citing' that the Secretariat had clarified that the nine-year moratorium agreed to in 2007 applied only to Botswana, Namibia, South Africa and Zimbabwe. A successful vote (49 in favour, 47 against, 46 abstentions) was taken on whether to open the debate. In the debate, Japan indicated that it felt that the downlisting would contribute to sustainable use, would improve the welfare of local communities and would support species conservation. The Congo, Rwanda and Sierra Leone opposed the proposal. A vote was held, by secret ballot at Zambia's request, and the proposal was rejected with 59 votes for, 47 against and 38 abstentions.

In the run-up to CoP 16, March 2013, only one proposal to trade in ivory was received – again from Tanzania, which sought permission for a one-off sale of just over 100 tonnes and with no further proposal to come from the country for a period of four years after the sale. Burkina Faso and Kenya submitted a proposal essentially reiterating the terms of the nine-year moratorium from 2007. Tanzania withdrew its proposal before the CoP began – Burkina Faso and Kenya withdrew theirs during CoP 16. Tanzania probably withdrew its proposal because the end of the moratorium period is not far off, and because in the context of significant increases in elephant poaching the proposal was unlikely to have been successful in 2013 anyway.

The 'decision-making mechanism' that it had been agreed at CoP 14 would be decided upon and adopted at CoP 16 proved not to be ready; but an 'action plan' was adopted to ensure the finalisation of such a mechanism for consideration at CoP 17. CoP 17 will be held in 2016 in Cape Town, and it is probably no coincidence that South Africa offered to host. Elephants will assuredly loom large on the agenda.

8 In Committee I, Tanzania split (by successful vote: 76:37:15) its proposal into two, one with the ivory sale aspect and one without. A secret ballot vote was then held on whether the proposal with the ivory sale aspect included could be submitted, and was rejected (59:60:13), whereas the proposal without the ivory sale aspect was approved by secret ballot vote (57:45:32). Tanzania then put the latter amended proposal forward to plenary.

It was also decided that the CITES Secretariat would, if it can obtain external funding, convene a task force[9] on ivory enforcement to review existing strategies, propose collaboration measures, consider research into ways to age and identify ivory and work on the training of relevant officials. In addition, the CITES Secretary-General is to write officially to the United Nations to express concern about illegal killing of elephants in Africa and the illegal ivory trade, the national security implications of such killing and trade for African countries, and to request that these be brought to the attention of the UN General Assembly and Security Council.

9 Comprising representatives from China, Kenya, Malaysia, the Philippines, South Africa, Tanzania, Thailand, Uganda, Vietnam and various relevant organisations.

6 The IWC, CITES and whaling

6.1 Downlisting proposals

The IWC's 1982 moratorium on commercial whaling might not have been successful without CITES on board. The 'introduction from the sea' of species in the Appendices constitutes international trade and needs a CITES permit, and introduction from the sea of species in Appendix I of the Convention is prohibited if for commercial purposes. From 1 January 1986, all cetacean species regulated by the IWC (bar West Greenland minkes) were added to Appendix I of CITES – this having been agreed to at CITES CoP 4, 1983, 'in order to bring the Convention in line with the IWC decision to halt commercial whaling from 1986' (Lyster, 1985: 36–7). It seems that there were early concerns about CITES covering the same species as other treaties, but that these were 'partially relieved' by the provision in Article XIV.4 that if Appendix II marine species are 'also protected under other treaties or instruments, and a state is party both to CITES and the other relevant treaty', that state does not have to abide by its obligations under CITES in so far as the particular species 'are taken by its flag ships and in conformity with the obligations of the other treaty' (Birnie, 1985: 394–5).

Birnie records that at IWC 29, 1977, the CITES Secretary-General, who had observer status, reported that 'agreement had been reached at CITES' first Meeting on categorisation of species and taxon by evaluating biological and trade statistics together'. These would apparently be classified under Appendix I if 'on biological grounds they were threatened with extinction and if, further, they were, or may be, affected by trade in them for any purpose'. Clearly, says Birnie, there will 'be scope for confusion if a close relationship' is not maintained. The IWC, it seems, 'agreed to offer its services to CITES as adviser on cetaceans and also adopted a resolution calling on IWC members to prevent import into their countries of whale products from non-member nations' (Birnie, 1985: 490–1). At a Special Meeting of the IWC in 1978 a 'Resolution to the CITES' was passed, as discussed earlier.

At CITES CoP 2, 1979, as already discussed, CITES adopted resolutions that CITES would consult the IWC (Res. 2.7) and that parties would not 'introduce from the sea' species protected from commercial whaling by the IWC (Res. 2.8). At CITES CoP 3, 1981, all cetaceans were listed on Appendix I or II, with Appendix I then including the blue, fin, gray, humpback, right, sei and sperm whales. This meant that this Appendix covered all whales listed at that date by the IWC as 'Protected Stocks' and more. All other species of cetacea, Birnie says,

> were placed on Appendix II, though Canada, Japan, Norway and the USSR made reservations on some stocks of sei and fin whales on Appendix I; and Canada entered a reservation for the Californian gray whale listings and Japan and Norway for the sperm whales.
>
> (1985: 576–7)

She then goes on to record that at CITES CoP 4, 1983, seven more cetacean species were listed on Appendix I, including some smaller cetaceans like the minke and pygmy right whales regulated by the IWC (ibid.).

I put it to Kleinschmidt that the resolutions dealing with the relationship between CITES and the IWC have been based on the premise that they have the same purpose – preservation, and then asked whether he agreed with this and how he saw the treaty's purposes. Kleinschmidt said:

> I think that it is in both instances about trade. In the case of the IWC, it is dressed up by the Japanese lobby as a statistical right; and it is dressed up by the others as not trusting those who want to do the fishing, killing.
>
> (Kleinschmidt, pers. comms, 2006, 2008)[1]

He then described CITES as being, 'in a way, a more modern instrument', and said that he thinks 'there is a point of tension because the one instrument was created for a broader purpose, and of course that broader purpose encompasses whaling as well, and to that extent there's an unresolved issue there'. Ultimately, though, he argued that 'one needs to see the greater context', which is that 'globally the oceans have no comprehensive and binding management systems', and where they exist they are for 'specific sectors such as over issues of whaling or over aspects of trade' (ibid.).

When I put it to Oosthuizen that the IWC has asked CITES not to undermine it – the wording being 'to recognise its authority' – but that proposals for the downlisting of whale populations have often been put forward at CITES meetings, he said that 'in the beginning' there had been 'some concern about what was going to happen at CITES', but that now people feel a two-thirds majority for downlisting will not be obtained, 'so they're quite relaxed about the role of CITES and so on'. He added, however, that 'if there was any indication that parties might get a two-thirds majority, then it would be different' (Oosthuizen, pers. comm., 2006).

1 To make this point more clearly, Kleinschmidt explained that, leaving aside historic or traditional rights to the oceans' whale populations, in his view, Iceland, Japan and Norway have 'at the backs of their minds' the intention to have dominant and preferential access to a meat source that, in time and according to their own calculations, will become an indispensable food source once other marine species are in even faster decline and the search for protein, on land and in the sea, escalates.

Donovan said that CITES is a trade organisation whereas, in his understanding, the IWC is not. Some parties, he said, dislike the fact that the moratorium is a blanket moratorium which does not take into account the status of individual whale populations – and that the moves to use CITES reflect this. Asked what effect approval of a downlisting of certain whale populations at CITES might have on the IWC, he said that 'in theory, it shouldn't have any effect, because they're not related issues', and that in practice it was really just 'sabre rattling' by opposing sides (Donovan, pers. comm., 2007). Asked about the relationship between the IWC and CITES, Grandy told me that she felt that the growing membership of the IWC reflected increasing environmental awareness as more states had become interested in the management of whales – which followed a trend for governments to be more interested in environmental issues in general. Asked whether it was possible to understand the treaties in isolation, Grandy said that she thought that the CITES Secretariat was actually trying to avoid undermining the conservation measures of the IWC, which she felt is quite normal behaviour for conventions (Grandy, pers. comm., 2007).

At CoP 10, Japan proposed to repeal the long-standing 1979 resolution (Res. Conf. 2.9) in which the CITES parties had supported the IWC by agreeing not to issue any import or export permit for any species protected from commercial whaling by the IWC. Russia argued that actions taken by the IWC should not determine decisions of CITES, and Burundi, Canada, Denmark, Iceland, Namibia and St Vincent all supported the draft resolution. The US, however, argued that the IWC has never withdrawn its request for support from CITES in enforcing the moratorium on commercial whaling and that CITES should not withdraw support from another international organisation which requested support. Specifically, Japan's draft resolution stated that the CoP:

> resolves that amendments of Appendices I and II with regard to whale species be decided under CITES' own criteria, taking into account (i) scientific information from the IWC together with other sources; and (ii) consistency with scientific relevance for the listing of other species in the appendices.
>
> (CITES, 1997)

The proposal was rejected by 51 votes to 27 in a secret ballot (CITES, 1997). This is a very important proposal, indicating, as it does, a deliberate effort by Japan to de-link the two treaties (paradoxically linking them politically) with the implication being that trade in whale products would become permissible under CITES.

A feature of CITES CoP 10 that did not receive the attention it perhaps deserved is that, at the final voting round of the Conference, a proposal by Norway to downlist certain minke whale populations from Appendix I received more votes in favour than against. However, it did not receive the two-thirds majority needed for a successful downlisting proposal to be adopted. The proposal did, however, attract more votes in its favour than there were at the time parties to the ICRW.

Komatsu and Misaki state that both Norway and Japan have put on record their objection to the listing, by CITES, 'of abundant species on a list of endangered creatures, and are therefore exempt from the prohibition' (2003: 106).[2] This is something common to elephants and whales – both (or at least certain sub-species of both) are non-endangered species being protected as though they were endangered. This is presumably another reason why they have become a 'battleground' – after all, if they were *truly* endangered it seems unlikely that the debate over their sustainable use could be so bitter.

It was argued by Japan and Norway that newly adopted criteria for listing on CITES should be followed, rather than continuing to follow the IWC (Gillespie, 2002). These criteria (the 'Everglades criteria', adopted by CITES CoP 9 in 1994) replaced the 1979 'Berne criteria' with which CITES had been operating, and provided for regular review of species' listings. In terms of the Everglades criteria, new considerations could apply – such as juxtaposing the positive and negative aspects of trade, in order to consider whether it might not be beneficial to the conservation of a species for trade to take place. This is quite a radical turnabout – which might even be considered as the reverse of the ICRW's 'Berlin Initiative' of 2003 – that trade might be *beneficial* to species.

Japan and Norway in fact submitted five separate proposals for downlisting specific whale stocks from Appendix I to II, which would have allowed international commercial trade. All of the votes were by secret ballot and none were adopted.[3]

In the lead up to the 1997 CITES CoP, the High North Alliance (HNA) NGO suggested that the importance of the five proposals was for the way in which CITES chose 'to relate to the IWC' and that it would potentially 'be decisive for the future of CITES'. By continuing to base its own decisions on the IWC's, the HNA argued, CITES would be 'endorsing a policy where myths, cultural preferences and the popularity of certain animals in urban public opinion, mean more than biological and ecological criteria and scientific assessment'. The HNA concluded that, 'as a result of such endorsement, CITES' credibility as a conservation body' would 'be severely impaired' (HNA, 1997). After the vote, CITES' credibility might in the HNA's view have been 'impaired', but the voting margins were narrow.

The result of the vote on the Japanese proposal for downlisting of Antarctic minke whales was 53 in favour, 59 against with four abstentions; and the Norwegian proposal for North Atlantic minke whales gained 57 in favour, 51 against with six abstentions. In summary, write Komatsu and Misaki, 'about half of the member nations voted in favour of downlisting the Antarctic minke whales, and more than half voted in favour of downlisting the North Atlantic minke whales' (2001: 152–4).

This was not the first time that the pro-whaling countries had tried to use CITES to circumvent the ICRW, and the voting shift was significant. When Norway proposed the downlisting for the first time in 1994, only 16 states voted in favour

2 By this, Komatsu and Misaki mean that Japan and Norway hold reservations to the CITES listings. Strangely, however, Japan does take sei whales in the Pacific Ocean despite not holding a CITES reservation in respect of these whales.

3 The proposals were in respect of the Eastern Pacific gray whale, the Okhotsk Sea-West Pacific minke whale, the Southern Hemisphere minke whale and the western North Pacific Bryde's whale (Japan); and the Northeast Atlantic and North Central Atlantic minke whale (Norway).

and 48 against. The ballot was then an open one, to which fact Komatsu and Misaki attach great significance. Since then, they suggest, CITES has moved to a secret ballot system, 'and the results of the more recent ballot reflect the wider and improved recognition of the need for rationale utilization of abundant whale stocks'. They conclude that the lesson to be drawn from this is that Japan 'is not in the minority in the international community that embraces developing nations' although Japan 'may be in the minority within the IWC which in reality is a forum comprising of a limited number of countries' (Komatsu and Misaki, 2001: 152–4). Obviously, this debate about ballot methods shows very different understandings about transparency and democracy.

Komatsu and Misaki consider that in these voting patterns there is 'considerable cause for optimism', as, they say, CITES represents a world-view considerably larger than the IWC. They then describe the Japanese as tending 'to look at the world through the eyes of developed nations, particularly through the eyes of the US and Europe', whereas a lesson which Japan could instead take from CITES is that this developed nation view is 'not the only view of the world, and that the global community is in fact significantly made up of developing nations, of which the Americans and the Europeans are just a part' (Komatsu and Misaki, 2003: 107). This is the view that the developing world is more oriented toward sustainable use ('conservation') than preservation – which is arguably true.

I asked Butterworth about the purposes of the ICRW and CITES, and also whether it is possible to understand either CITES or the IWC in isolation from each other. In this regard, I posited it as an irony that CITES was once more protective than the IWC, but probably no longer is, and asked how significant it might be, given this swop of positions, that CITES is a far more representative body (numerically) of the world's states. Butterworth suggested in reply that the swopping of positions was 'a trend which has probably stabilised now' and that the CITES option had been pursued 'because the whaling countries in 1994 got a fair amount of support for the downlisting of minke whales', and because 'the IWC was deadlocked and this was seen as a way of making a moral case argument'. He argued, however, that the 'game' has 'probably been played to a conclusion by the EU operating as a bloc', and because of the 'control at the moment on issues like this by NGO movements'. These factors, he argued, 'probably mean that there's no chance that CITES could ever go to a majority on this' – but that the use of CITES 'served a purpose, to highlight the issue' and 'also to show that CITES on this issue is not working in terms of its criteria' (Butterworth, pers. comm., 2006).

I have argued that Japan and Norway did very clearly attempt for some years to play off the IWC and CITES against each other, and in fact attempt even to undercut the IWC through attempts to downlist whale species at CITES. On the flip side, of course, it can be charged that the anti-whaling states are essentially doing the same thing by using their votes at CITES to uphold IWC resolutions, an arrangement which CITES agreed to before Japan had even become a member; although in recent years this seems to be less of a deliberate policy than a reaction to the Japanese and Norwegian efforts. After all, it is Japan and Norway which put forward the downlisting proposals at CITES CoPs.

In 1998, at IWC 50, a resolution on 'Cooperation between the IWC and CITES' was passed. Both Japan and Norway opposed the resolution, arguing that it was 'trespassing on the competence of other organisations' – such other organisations ('more appropriate forums') being either the WTO or CITES (Gillespie, 2002: 27). Despite the export of whale products being a trade issue, it is hard to support an argument that the WTO would be a more appropriate forum than the IWC for dealing with issues relating to cetacean conservation, or even that the WTO would be a more appropriate body than CITES.

In 1999, at IWC 51, the IWC agreed to a resolution recognising that it has not yet completed the necessary measures to ensure that commercial whaling catch limits are not exceeded, that whale stocks can be adequately protected and that all whaling by IWC member countries is brought under effective IWC monitoring and control (IWC, 1999). With this resolution, the IWC reaffirmed the importance of its relationship with CITES and advised CITES once again that zero catch limits are still in force for all species of whales managed by the IWC. By the time of the 2001 IWC meeting, the IWC (or at least the majority of its contracting governments) clearly felt the need, however, to entrench its position. In a resolution (the 'Resolution on Interactions between Whale and Fish Stocks'), the IWC described itself as 'the universally recognised international organisation with competence for the management of whale stocks' (IWC, 2001).

An important thing to note about the proposals to downlist whales, which were made at CITES CoPs in all of 1994, 1997 and 2000, of course, is that support for them is at least constant if not increasing. It would be difficult to ascribe such a high degree of support to bribery by, or pressure from, the pro-whaling states. Komatsu and Misaki offer changing mores as an explanation, writing that the 'problem of whales' is now recognised as something that 'should not be addressed solely by the IWC', and that:

> when we look at the ocean system of the world as a whole, it is inevitable that whales must be studied from the perspectives of multi-species management because of their position at the top of the ocean food web and because of their wide and long ranging migration.

Times, they suggest, have changed 'since whales were considered as species independent from other components of the oceans. Now, the management of whales must be considered in the context of various factors with fisheries in mind' (Komatsu and Misaki, 2003: 154–5).

6.2 Further downlisting proposals

At CITES CoP 11, in 2000, Norway pushed again for a downlisting of minke whales. In the lead-up to the CoP, the HNA pointed out that Resolution 2.9 from 1979 does not oblige CITES to follow the IWC; rather, that the resolution 'only recommends the parties not to issue any international trade permit for whales protected by the IWC' (HNA, 2000: 5). The HNA also pointed out that CITES did not, in the 1980s, follow IWC decisions automatically. The example chosen to illustrate this point is, however, ironic – given that CITES afforded *greater*, not less, protection to species such as Bryde's, fin and sei whales (listing them on Appendix I even while the IWC was still issuing quotas). The HNA uses a better example for its purpose when pointing out that in 1983 CITES kept the West-Greenland minke whale on Appendix II, on the basis that the species was not subject to an aboriginal subsistence hunt (ibid.).

After meeting opposition early on, on the last day of the CoP Norway called for a re-opening of the debate and amended its proposal with further DNA-based identification strictures to ensure that trade was undertaken only with countries where such identification systems were implemented. Several delegations opposed the proposal, arguing that downlisting the minke whale would amount to resuming commercial whaling. The proposal was voted on in secret ballot and again failed to reach the required two-thirds majority for downlisting. Once more, however, votes in favour outnumbered votes against – with 53 votes in favour, 52 against and eight abstentions (CITES, 2000).

In 2001, at IWC 53, Resolution 2001-5 (the 'Resolution on Commercial Whaling') expressed concern that:

> the government of Norway, having lodged a reservation to the CITES Appendix I listing of whales, has announced its intention to resume international trade in minke whale products despite the decision by the CITES Conference of Parties in 2000 to support the continued listing of minke whales on Appendix I.

The resolution 'requested' that the Norwegian government issue no export permits for whale products, 'called upon' the same government to 'halt immediately all whaling activities under its jurisdiction', and 'instructed' that a copy of the resolution be forwarded to the CITES Secretariat.[4]

CITES CoP 12, 2002, saw the whaling debate resume. Japan officially submitted proposals to the CITES Secretariat for changes to the classification of Northern hemisphere minke whales and Northern Pacific Bryde's whales. Japan proposed downlisting these species from Appendix I to Appendix II. Voted on, the proposal to downlist minke whales failed with 41 votes in favour, 54 against, five abstentions and six spoiled votes. The Bryde's whale proposal failed also with 43 votes in favour, 63 against, three abstentions and two spoiled votes.

The specific allegation was even made that Japan had been trying to undermine the IWC with its motions at CITES. In the anti-whaling NGO publication, *Eco News*, Kitty Block, for the Humane Society of the US, was quoted as saying that 'Japan was unable to circumvent the IWC whaling moratorium by playing one

4 Resolution 2001-5 also referred to Resolutions 1994-7, 1995-6, 1996-3, 1997-2, 1998-8 and 1999-6 which were all 'welcoming' of 'the continuing cooperation between CITES and the IWC on issues related to trade in whale products'.

Convention off another' (*Eco News*, 2002). It is strange, though, that – with the occasional exception – the linkage between the two treaties has gone largely unremarked upon by international commentators, and seems to have escaped serious analysis almost completely. It is, of course, not a particularly dramatic point – at least, not one that mainstream media could realistically be expected to seize on.

It was, however, reported from the CoP that in the plenary session, the matter of CITES's cooperation with the IWC was raised. Iceland argued that 'CITES principles are breached by any state which opposes resumption under any circumstances of commercial whaling', and Antigua and Barbuda stressed that 'cooperation with the IWC should be based on sustainable international trade' (*IISD*, 2002).

Along with the downlisting proposals it put forward to CoP 12, Japan presented to the CITES Secretariat two draft resolutions dealing with the relationships between CITES and the IWC, and between CITES and the FAO. In the first resolution, it seems that Japan called for changes to be made to CITES Appendices (insofar as they apply to cetaceans) 'only on the basis of scientific information and for controlled trade in whale products to be allowed between IWC member states'. In the second resolution, Japan called for 'the views on CITES Appendix listing criteria [*sic*] of the FAO and regional fisheries management organisations to be given greater respect' (CITES, 2002).

Could CITES replace the IWC, at least to an extent? This seems unlikely. Burns and Wandesforde-Smith comment that despite all of its threats to withdraw, Japan has not done so and they conclude that the ICRW 'is likely to remain the focal point for the management of commercial whaling in the future' (2002: 210).

From CITES's side, too, there arguably is a concern about being undermined, or, at least, a desire to prevent CITES from being used to circumvent the IWC. At the end of CoP 12, a resolution on the relationship between CITES and the IWC was passed, recording that:

> the parties agree not to issue any import or export permit, or certificate for introduction from the sea, under this Convention for primarily commercial purposes for any specimen of a species or stock protected from commercial whaling by the [ICRW].
>
> (CITES, 2002)

The preamble to the resolution recorded that the parties:

> CONSIDER that the International Whaling Commission has asked for the support of the parties in protecting certain stocks and species of whales; . . . ARE MINDFUL of the need for special attention to the conservation of whales and other cetaceans; . . . RECALL also that the great whales have not generally recovered from the depletion brought about by commercial exploitation, even though many other species of exploited wildlife have recovered from equal or greater degrees of depletion; . . . [and] AFFIRM [their] concern that any

illegal trade in Appendix I whale specimens undermines the effectiveness of both the IWC and CITES.

The whaling issue, it has been suggested by Stoett, 'strongly captures the inherent complexity of the political struggles to control or manage the global commons, especially the human competition and cooperation resulting from the controversy that so often ensues when territorial boundaries are unclear' (1997: viii). The African elephant, I think, epitomises similar complexity and the fates of the two species are increasingly bound together in international law and politics.

According to Komatsu, in the present circumstances, 'we must achieve support from other organizations one by one until the IWC becomes isolated in the international community'. Toward this end, they explain, 'we need to take a multi-facet approach to the opening of the new sustainable whaling' and 'we chose CITES as our first approach to other international organizations' (Komatsu and Misaki, 2001: 151–2). This is crucial, and it arguably blows away the charges of naïvete levelled at the Japanese (unless one argues for the word in a different sense, and contends that by making the links so specific the Japanese have been clumsy negotiators). In a sense, too, this is the 'smoking gun' for efforts to link the species and treaties. Komatsu[5] is actually suggesting using CITES to undercut the position of the anti-whaling majority at the IWC. Further, he is recording that this was a policy of Japan's.

Joji Morishita told me that Japan felt that there was a connexion between the two treaties as long as CITES used the IWC list as a basis for its own listings, but that Japan's proposals to CITES had been based on CITES's own criteria (Morishita, pers. comm., 2011). Goodman agreed and told me that Japan's proposals within CITES were 'very directly affiliated to CITES itself' (Goodman, pers. comm., 2011). Morishita added that 'the IWC from our point of view has no jurisdiction or competency over international trade, while CITES has only jurisdiction over trade' (Morishita, pers. comm., 2011).

At the 13th CITES CoP, in 2004, Japan put forward a proposal that CITES 'urge the IWC for an expedited completion of the RMS', with the proposal being 'aimed at enabling CITES to make decisions on trade regulation of whales in accordance with its Appendixes expeditiously' (*JWA News*, 2005b). After argument, the proposal was voted on and failed, but fairly narrowly, with 57 votes for, 63 against and 13 abstentions. After this, Japan put forward a proposal to downlist minke whales from Appendix I to II; specifically, downlisting of the Northern hemisphere minke whale, based on the fact that its 'present abundant state' does not comply with the requirements of Appendix I. Voted on, the proposal failed with 55 for, 67 against and 14 abstentions.

Strangely, however, *JWA News* argued that at the same CoP, 'on other fronts', Japanese fisheries sources had 'expressed a concern about an apparent trend in

5 Represented Japan at CITES, FAO and IWC. Ceased to represent Japan at the IWC not long after the events in Shimonoseki in 2002.

CITES toward its greater involvement in the management of marine species, which, they see, would call for adequate caution from the viewpoint of sustainable users' (*JWA News*, 2005b). This implies that Japan is probably not in favour of the wholesale linking of treaties, but that it wishes to see this happen selectively. The further implication is that the downlisting proposals were indeed being used to put pressure on the IWC, rather than being intended to presage a new era of different management.

6.3 A change of approach

In the lead-up to the 14th CoP in 2007, Japan submitted a Working Document on 'interpretation and implementation of the Convention' in which it argued that the 'scientific veracity' of the listing (between 1975 and 1986) on Appendix I of all species of whales managed by the IWC has 'not been tested against the most recently adopted criteria for listing species in Appendix I'. The Document then recalled that in July 2000, the Secretary-General of CITES had written to the IWC and 'expressed serious concern' about the 'escalating and increasingly divisive conflict' within the IWC, before expressing the view that it was 'therefore crucial' that the IWC 'soon make important progress toward the adoption of a Revised Management Scheme' which would enable the CITES CoP to 'adopt the appropriate management regime' for listed species of whales. The Document recalled that 14 separate proposals to downlist species of whales had been put forward at the last four CITES CoPs, all of which were 'denied' by the CoP. Japan's Document pointed out that parties opposing the transfer proposals have 'maintained that CITES should not transfer these species before' the IWC's RMP and RMS were completed, but also pointed out that at IWC 58, in 2006, the IWC 'agreed to postpone any further discussion towards completion of an RMS'. Japan's Document then argued that 'appropriate listing of cetacean species on the Appendices of CITES would ensure proper international regulation of the trade', and suggested that CITES's 2005 Strategic Vision, Goal 2, included two objectives relevant here: first, that it should be ensured that CITES's Appendices 'correctly reflect the conservation and management needs of species' and second, that it should be ensured that 'decisions to amend CITES' Appendices are founded on sound and relevant scientific information and meet the agreed biological and trade criteria for such amendments' (CITES, 2007).

The Working Document then proposed two draft decisions. The first, directed to the Animals Committee of CITES, proposed that the Committee:

> shall include in its Review of the Appendices all cetaceans in Appendix I that are managed by the IWC. In addition to evaluating the listing of these species against the criteria for inclusion of species in Appendix I contained in Res. Conf. 9.24 (Rev. CoP13), the Animals Committee shall, for consideration at CoP 15, propose amendments to update Res. Conf. 11.4 (Rev. CoP12) based on: i) a review of the current situation regarding illegal trade in whale products, ii) advice from the IWC Scientific Committee concerning the status of whale

> stocks and, iii) the decision of the IWC that the moratorium on commercial whaling is no longer required.
>
> (CITES, 2007)

The second draft decision was in the form of a Direction to the CITES Secretariat to write to the IWC Secretariat:

> conveying the concern of the CoP regarding the postponement of the RMS discussions and, in the spirit of Article XV, paragraph 2.(b), to ensure co-ordination with conservation measures, request scientific data and seek advice concerning the listing of whale species on CITES Appendices in the absence of any prospect for the completion of the RMS.
>
> (CITES, 2007)

Right at the start, in a discussion of the Report by the Animals Committee, Argentina, supported by Germany on behalf of the European Union, argued that no review of whale stocks ought to take place under CITES, as the IWC was examining whale stocks. The matter was referred for Committee discussion.

In Committee I, in proposing the periodic review, Japan argued that it would ensure that CITES 'operates on the basis of current scientific information', and that the proposed review would have no effect on the IWC moratorium. In support, both Norway and St Kitts and Nevis warned that decisions within CITES ought not to be based on any criteria but scientific ones; and China stated that it would welcome any additional information which would result from the proposed review. In opposition, Australia argued that the IWC is the 'agreed competent authority'; Argentina argued that the proposed review would duplicate the processes of the IWC's Scientific Committee; and the European Union reminded parties that the IWC had recently passed a resolution on interaction between CITES and the IWC in which it is stated that the moratorium on commercial whaling is still in place. It was decided by vote – 54 against 26 – that great whales will not be subjected to a periodic review for so long as the moratorium in the IWC remains in place. Another vote was taken on a proposal by Australia that no periodic review of any of the great whales should occur whilst the IWC's moratorium remains in place, and this passed by 59 votes to 21. (This proposal had been in response to a proposal from the Animals Committee (AC) to include in the periodic review the central stock of the North Atlantic fin whale (*IISD*, 2007).)

Late in the CoP, Palau apparently did try to revisit the periodic review of whales. What seems to have happened was that Palau failed by one vote to re-open debate (35 for and 71 against, with one-third support being needed). Plenary therefore adopted Committee I's decision, and the CoP instructed the AC that 'no periodic review of any great whale, including the fin whale, should occur while the IWC moratorium is in place' (*IISD*, 2007).

Japan has reservations lodged in respect of the common (Northern) minke whale (except for the Appendix II-listed population of West Greenland), the Antarctic minke whale, the sei whale (although not to populations in the North Pacific, despite taking such under scientific permits, and between the equator and the Antarctic), the Bryde's whale, the fin whale, the sperm whale, the Irrawaddy dolphin and the Baird's beaked whale, which are all listed on Appendix I. Iceland has reservations over the common (Northern) minke whale (except for the Appendix II-listed population of West Greenland), the Antarctic minke whale, the sei whale (unrestricted), the blue whale, the fin whale, the humpback whale, the sperm whale and the northern bottlenose whale. Iceland also has reservations lodged over various Appendix II cetaceans, these being the West Greenland population of the humpback whale, seven species of dolphin, the killer whale and one species of porpoise. Norway has reservations lodged over the common (Northern) minke whale (except for the Appendix II-listed population of West Greenland), the Antarctic minke whale, the sei whale (although not to populations in the North Pacific and between the equator and the Antarctic), the fin whale (although only in respect of populations in the North Atlantic off Iceland and Newfoundland, and in a southerly region near Antarctica) and the sperm whale. Palau has reservations lodged in respect of the common (Northern) minke whale (except for the Appendix II-listed population of West Greenland) and the sperm whale. St Vincent and the Grenadines has a reservation lodged over the humpback whale.

Sand argues that the IWC does not take priority over CITES in respect of whales, but that the two autonomous global regimes supplement and complement each other. He describes this as being 'a variety of interplay not untypical in modern international law' (Sand, 2008: 60–1). In this chapter, the case has been made that certain pro-whaling states have deliberately tried to shift the debate from the IWC – which they may have seen as being deadlocked, and possibly beyond salvation – to CITES. In response, anti-whaling states have tried to ensure that CITES continues to defer to the IWC. At the same time, it is important to understand the extent to which the debate over how best to utilise wildlife generally has shifted – with terminology changing and new understandings of the complexity of ecosystems being argued for. This has made the debate far more complicated than a mere choice between 'using' and 'not using'. The argument that there are 'too many whales' and that humans need to use them, mirrors the argument that there are 'too many elephants' and that elephants must 'contribute to their own survival'. As Morishita said to me in 2011, 'discussions about elephants and whales are quite similar – you can just exchange the name whale and elephant and you might not realise we are talking about different animals' (Morishita, pers. comm., 2011).

While in one sense my argument (at its narrowest focus) is complete, at this point it is important to turn to an examination of various related aspects – which all throw light on different angles of the issue, and without an understanding of which a full understanding of the argument is not possible. What also needs to be considered is the global importance of the political and legal linkages established between the two treaties and the two species.

7 Conservation of elephants and whales

7.1 Conservation's 'poster children'

The elephant issue highlights two different approaches that can be taken to conservation. The 'Kenyan model' attempts to maintain elephant populations by ensuring absolute protection and by banning all trade in elephant products. The approach of the Southern African countries is to manage elephant populations actively, occasionally culling animals to provide sufficient habitat, and (ideally) sharing profits with local people and ploughing profits back into conservation. There is a wide gap between different camps' understanding of and appreciation for elephants. In Parker's forceful words, 'elephants as sources of pure joy to Africa's visiting tourist are worse than smallpox to the peasant whose crops they have just destroyed' (2004: 381). Leakey, however, suggests that the goal in Kenya is 'clean air, clean water, plentiful forests, and a human population that is well fed, educated, and reasonably affluent' and that 'saving the elephants is symbolic – a means to achieve these greater objectives' (2001: xi). This is a different angle to the elephant as a symbol – not just a symbol for preserving wildlife, but a symbol for improving human life – for sustainable development itself.

Kreuter and Simmons, quoting Western, write that 'the African elephant is an important conservation *symbol*', and that it has been claimed to be an important 'conservation flagship' because 'it evokes strong sympathy and can, given public support, protect the integrity and diversity of African ecosystems'. The same writers, quoting Peterson, note that the controversy over elephant conservation is 'a controversy that is less a rational discourse than a conflict between widely differing socio-economic and symbol systems' which 'has evolved due to two diametrically opposed perceptions of the relationship between mankind and nature embodied in the preservationist philosophy, where mankind is excluded from nature, and the utilitarian philosophy, in which mankind has dominion over nature' (Kreuter and Simmons, 1994: 39, 40). This is the basic debate.

Kreuter and Simmons then write that Western conservationists have created a 'mediagenic conservation totem through which *Elephant extermination imminent* is reported to have become the most lucrative slogan in the environmental movement's history' (1994: 49). An important link between species is that the same point has been made about the whale – the 'super whale' (Kalland, 1994: 163). The whale exemplifies the same divergence of opinion as does the elephant.

To whaling nations, writes Bowman, 'the relatively plentiful stocks of certain cetacean species has meant that there could be no legitimate justification for refusing to sanction the resumption of commercial whaling'; for others the 'real need' is to prevent both the extinction of cetacean species and the death and suffering of 'individual sentient and highly intelligent creatures' (1995: 5). To other commentators, like McGuinness, once the 'rational aim of ensuring the survival of species has been achieved', he says, the rest becomes merely 'a hotchpotch of emotion and superstition. Of course whales are remarkable beasts, but so are pigs – and no less intelligent. What makes whales morally superior to pigs?'. His answer, in the Australian context, is that it is bureaucrats who are pushing for the preservation of whales – in response to middle class concerns (McGuinness, 2002). While the writer's language here is as emotive as is that of the 'whale preservationists', an important point is raised in the suggestion that the distinction between whales and pigs, or any other animals, may indeed be an arbitrary one that we need to move beyond.

It might well be that peoples' and nations' views, globally, *are* changing. Komatsu and Misaki record that in 1992 only five nations – Iceland, Japan, Norway, St Lucia and St Vincent – formed the 'pro-whaling bloc', but 'now we look around and find more countries are on our side'. They conclude that Japan feels 'more comfortable now than we did in the earlier days', and that this 'is because we know that our cause for the sustainable use of marine resources is gaining wider recognition' (Komatsu and Misaki, 2001: 160–1).

Essentially, there are two ways to view the sustainable use of natural resources – consumptive and non-consumptive. The split in the usage of both elephants and whales mirrors this – there are those states which argue for non-consumptive use, and those which argue that the money which can be made from consumptive use can be ploughed back into conservation and, in this way, wildlife pays for its own protection. While the battles over how best to use and manage elephants and whales continue to rage, and as opposed views seem ever less compatible, the two species have gradually been turned into icons of the cause – with the exponents of each view pushing their 'poster child' forward as a symbol.

7.2 African viewpoints: Eastern and Southern

According to Daphne Sheldrick, 'any trade' in ivory threatens all elephants as it is 'no good thinking of them in pockets. Elephants can walk 100 kilometres in a day. What happens at one end of Africa impacts on the other' and there is 'an upsurge in poaching every time there is a CITES meeting'. Of the argument that the money raised from a legal trade in ivory can be ploughed back into conservation, she suggested that 'history shows that it is coming back into pockets'. This is arguably an unfortunate reality behind arguments that conservation *needs* trade. Despite its various problems, Kenya, Sheldrick suggests, has much to be proud of; although, she told me, the country 'is having a hard time in the world, defending its stance'. Questioned on Tanzania's apparent switching of stances, she said cynically that 'money talks', but that 'Kenya has always been known for its compassionate stance' and that this is of 'great market value to Kenya' (Sheldrick, pers. comm., 2004).

Douglas-Hamilton explains that people in conservation circles in East Africa are convinced that 'a resumption of the ivory trade in any form would lead to a resumption of the uncontrolled illegal trading of the past' (1997: 20). Without trade, the stockpiles of ivory kept by governments will inevitably increase and pressure will always build to trade them; but this has to be weighed against the dangerous consequences of any trade setting a precedent. Douglas-Hamilton, worth quoting at length on this, writes that it is a powerful argument to trade ivory and use the proceeds to finance conservation schemes which might benefit local people sharing the land with elephants. 'Unfortunately', he argues, 'the argument, while compelling, is fatally flawed' because:

> all past experience suggests that any legal trade will stimulate a vast, illegal, parallel trade both in Africa and at the consumption end in Asia. Secondly, it is questionable that the proceeds from ivory would benefit conservation or the local people, rather than disappearing into the pockets of power brokers.

'Finally', he concludes, 'any legal trade would undermine the present climate of disapproval associated with owning ivory, sending out a clear signal that ivory is respectable provided it comes from official sales' (ibid.).

In South Africa, particularly, it can be argued by the protagonists of ploughing the proceeds of trade back into conservation that government corruption will not see the proceeds diverted into the pockets of politicians and the wealthy elite – as must be a grave concern in present-day Zimbabwe. There is, however, another form of 'corruption' that has taken firm root in South Africa in recent years – the 'corporatisation' of parastatals. In the conservation field – as in municipalities, universities and parastatals generally – the move toward corporatisation, the creating of executive structures with 'executive salaries' and an ever-widening gap between these structures and other staff, has become a feature. While these executives are appointed often on short contracts, without long-term job security, the (seemingly almost irresistible) temptation is there to engage in 'corporate looting' – especially through the iniquitous medium of often undisclosed, self-assessed and ludicrously high 'performance bonuses'. To my mind it makes little difference whether the 'proceeds going back to conservation' find their way unofficially (secretly) into the pockets of middlemen and corrupt officials, or whether they find their way into the grossly inflated official salaries of the executive management of the conservation hierarchies. According to the 2005/6 annual report released by South African National Parks (SANParks) in South Africa, the salary package for the executive team of 12, including bonuses and incentives, was 30 per cent higher than in 2004/5 (Bamford, 2006).[1] The cynical observer might well argue that in such a

1 It was reported in October 2006, for instance, that the SANParks spokesperson was earning almost as much as the president of the country's official salary. Including allowances, bonuses and a basic salary, this communications official earned considerably more than the national Minister of Environmental Affairs, under which Ministry SANParks falls. According to the same report, the whole SANParks management team earns 'seven-figure salaries' (Bamford, 2006).

culture of 'corporate greed', conservation considerations will be secondary to the finding of new ways to maximise profits; while it is also understandable that managers at lower levels, faced with budget constraints caused at least partially by the need to accommodate executive packages, will likewise argue for the increased use of resources such as ivory. In a world where the use of ivory has traditionally been politicised and where economic considerations are paramount, this is hardly surprising.

In mid-2007, just before the CITES CoP, I asked van Schalkwyk what role, given the opposition by Kenya and West African countries in particular to the ivory trade, he thought South Africa might be able to play in CITES in bringing African countries together. He answered that it is 'a very important foreign policy priority for South Africa to get greater African unity, greater consultation in Africa on issues of importance'. He then explained that:

> on this issue, we've tried our level best to discuss it with our African counterparts, but there is a difference here and the difference is, to put it quite bluntly, we have managed our wildlife in SA extremely well; some other parts of Africa did not, and we cannot be punished because they didn't manage their resources well.
>
> (van Schalkwyk, pers. comm., 2007)

It has been suggested that national parks are in trouble in East Africa, and that they are not bringing in the money they could, and were once expected to, bring in. Komatsu and Misaki write that 'the revenue they previously generated is now drying up'. With fewer tourists, they suggest, 'coming to the national parks and the resultant fall in income, the parks themselves are becoming dilapidated through lack of maintenance'. 'An additional problem', according to the same authors,

> is that many of the wild animals in the parks are overprotected to the extent that now there are simply too many of them, and they are becoming difficult to control. Elephants are a case in point, with protection programs having been too successful and Kenyan elephants now having to be transported to new habitats in West Africa, such as Sierra Leone and Cameroon.
>
> (Komatsu and Misaki, 2003: 150–1)

Probably, these charges are not entirely correct and are not to be taken too seriously – while tourism may have lost some lustre in recent years, it remains a significant earner of foreign currency. Further, if any elephants have been moved to the countries mentioned, it was in very small numbers and for reasons other than overpopulation. The comments are useful, however, for revealing something of Japanese awareness of linkages.

Komatsu and Misaki describe the 1989 Kenyan burning of ivory as 'an example of an irrational approach to the responsible utilization of wildlife' which 'might have made some people feel better, but did little to open minds to the real issue'. The writers then pose the question of:

> why the Kenyans do not want to responsibly utilize their own wildlife resources. If elephants are increasing to such an extent that they are now a problem, why don't people make use of them and cull the population to a sustainable level? What is wrong with a cull of set numbers, dictated by responsible conservation standards, and the people of Kenya using the meat and ivory and processing the skin and other parts that have commercial value?

The writers do not, however, make any effort to consider the Kenyan point of view – in other words, they do not present the Kenyans' arguments as to why they are not following this suggested, and apparently self-obvious, path. 'Unfortunately', Komatsu and Misaki continue, 'Kenya appears to have been brainwashed by a one-eyed protectionist policy that refuses to allow rational debate on the proper use of natural resources'. Their assessment is that Kenya has 'fallen hostage to the protectionist doctrine that killing elephants, at any level, or in any way, shape or form, will lead to the decline in its tourist industry' (Komatsu and Misaki, 2003: 150–1, 151, 152). This is much the same sort of rhetoric as is used in the whaling debate, and it creates a straw man of the Kenyan argument by oversimplifying it. This is a disservice to the East African viewpoint; a viewpoint which is also shared by elephant range state India, many Western states and many conservation groups.

Certainly, there appears to be an enormous gulf between the Eastern and Southern African views. Pickover writes, of an interview with the CITES coordinator in the Kenya Wildlife Service, that the interview 'highlighted the stark policy differences between Southern African and other African countries'. She quotes her interviewee as saying that 'one day we will realise that elephants cannot be reduced to the value of their teeth. Elephants are and always will be synonymous with the greatness of Africa' (Pickover, 2005: 77).

Western, however, told me that there is a wide split within East Africa itself, suggesting that 'the rift between Kenya and Tanzania alone is huge' (pers. comm., 2004). Western is not afraid to criticise Kenya for having policies that are contradictory, or at least inconsistent. At the same time, he also criticises the Southern African countries, arguing that they are not able to control poaching and that their arguments are not sound. To me, he described Kenya as having a contradictory attitude to sovereignty, as saying that elephants are 'common heritage', but then 'turning around and not liking being told what to do'. Our wildlife laws, he says, 'are very clear'; wildlife 'must pay its way outside of parks – that's our legislation' (Western, pers. comm., 2004).

Western argues that CITES has, or may have, reached its limits, and that it is not effective as a conservation tool within countries. He contends that part of what this says is that 'our conservation awareness is way outstripping these conventions, which were set up in previous periods and for different uses'. Asked about awareness, Western felt that this is poor. People on the ground understand little of CITES, and state parties might have little understanding of each others' positions. Most people, he says, are largely ignorant of CITES, with the only perception being that it has brought a stop to trade in ivory and rhino horn. This is a problem as it means that only a small percentage of Kenya knows about CITES.

There would be, he says, a lot more awareness of it here if there was hunting, but he has even 'heard one of our ministers blame lack of hunting on CITES'. South Africa, he concludes, 'has misunderstood the Kenyan position', where, of half a billion US dollars, 80 per cent can be attributed to wildlife – 8 per cent of Kenya's GDP. Hunting, he says, contributes nothing, and the more extreme groups have hammered the harm to the tourism industry (Western, pers. comm., 2004).

Lapointe has written that 'Kenya is free to choose how it resolves such issues for its people', but that other nations 'look to the surplus of their wildlife resources to help eliminate poverty and fund resource management programs'. 'What has been demonstrated', he continues, is that Kenya 'has voluntarily broken the established dialogue and has acted, not in the interest of the range states, but out of its own self-interest, an interest that curiously coincides with NGOs opposed to any sustainable use of natural resources' (Lapointe, 2003: 104, 129–32).

In Nairobi National Park, outside Nairobi, Kenya, there is a monument which is the ash pile at the spot where the ivory stockpile was famously burned in 1989. There is also a large marble plinth with a small sculpture of three elephant. A brass plaque on the side of the plinth reads:

> GREAT OBJECTIVES OFTEN REQUIRE GREAT SACRIFICES. I NOW CALL UPON THE PEOPLE OF THE WORLD TO JOIN US IN KENYA BY ELIMINATING THE TRADE IN IVORY ONCE AND FOR ALL. H. E. PRESIDENT DANIEL T. ARAP MOI C.G.H., M.P.

The coat of arms of Kenya appears above the writing. On another side of the plinth is a plaque which reads:

> THIS MONUMENT WHICH COMMEMORATES THE BURNING OF 12 TONS OF IVORY BY H. E. PRESIDENT DANIEL T. ARAP MOI C.G.H., M.E. ON JULY 18TH 1989 WAS MADE POSSIBLE BY THE GENEROSITY OF THE EAST AFRICAN WILDLIFE SOCIETY AND THE WORLD WIDE FUND FOR NATURE.

The logos of the two societies mentioned appear above the writing. Also at the site of the monument is an information board which, reproduced in full, reads:

> A Nation Stands Up For Its ELEPHANT In a period spanning only 20 years, poacher gangs traversed Kenya's rangelands and forests slaughtering rhino and elephant almost to the brink of extinction. Elephant numbers would decline from a high of 130,000 in 1973 to a paltry 16,000 in 1989. An insatiable demand for ivory fueled this slaughter, with a kilo going for as much as 13 US dollars. In 1989, poachers were killing an estimated 1000 elephants a month. It was time to put the foot down . . . The Wildlife Department's ammunitions store at the time was full with 2000 tusks meant for sale with bids stretching up to US $3.2 million. And that is when Dr. Richard Leakey, KWS Director, suggested that it all be torched in symbolic outrage at the

international trade in ivory. But would the ivory – made up of hard dentine and enamel – burn? And more significant, would the Government of Kenya agree to the burning? Luckily, the idea received the support of President Moi and it turned out that the ivory would burn fiercely if doused in petroleum. Burn Ivory . . . Ban! On the morning of 18th July 1989, Kenya's Head of State lead his country in making a statement of global concern: that Kenya would no longer allow the slaughter of its elephants to satisfy market demand for ivory. On this very spot, he faced a pile of ivory worth an estimated 1 million US dollars and lit a bonfire that was broadcast by television stations across the world. The publicity arising from this historic event stunned the world. The Convention on International Trade in Endangered Species (CITIES) (*sic*) placed the elephant on Appendix 1 – in effect enforcing a total ban on trade in ivory. The once lucrative ivory market collapsed, with price per kilo dropping to 1.50 US dollars. In the meantime, a rejuvenated and better-trained KWS ranger force routed poacher elements from Parks. Elephants still face the threat of poaching. In addition, their habitat has declined; their migratory routes are getting closed up as a result of settlement and there exists a fierce competition for resources, which brings the Jumbos into conflict with man. Do Not Collect This Ash! It must remain, as a symbol of this country's commitment to elephant conservation. And as you picnic here, reflect and join Kenyans in saying "Never Again".

(pers. photo., 2004)

South Africa and Kenya are arguably the two African countries best-known for wildlife tourism. Wildlife management would be well-served by rapprochement between the two, and the two countries may not really be as far apart as they presently seem to be. They reached a compromise on ivory at CITES CoP 14, and are the only two African countries which have historically supported the anti-whaling bloc at the ICRW.

At IWC 54, in 2002, Kenya stated that, although it had not attended for some time, it had 'nevertheless followed keenly the work of the Commission'. Kenya then referred to 'its government's strong commitment to the conservation of whales', and 'stressed the importance of tourism and therefore its eco-system to its economy'. It 'noted' that it had been the first African country to 'establish a marine sanctuary' and that 'non-consumptive use' was a policy that had 'served it well in both its cultural and economic development' (IWC 54, Chair's Report). On whales, then, Kenya and South Africa appear to be ad idem – while all other African IWC members vote almost exclusively along the pro-whaling line.

7.3 South Africa and conservation

South Africa ceased whaling in 1975, by which time its whale fishery was no longer of much significance. The country abstained on the moratorium vote in 1982 and since then has been generally on the side of the anti-whaling countries.

In mid-2006, I interviewed Maria Mbengashe who suggested that 'South Africa's general approach *is* sustainable use', whilst the country remains 'conscious of conservation' and taking 'consideration of benefits'. She described this as a 'new approach', saying that South Africa's 'approach *is* changing' with the 'emphasis changing toward sustainable use and development'. Questioned as to how whaling fits into the 'sustainable use' approach, Mbengashe explained that 'nonconsumptive use benefits South Africa in terms of tourism and generation of income' (pers. comm., 2006).

Questioned as to why South Africa would be against Japan making consumptive use of whales, Mbengashe said that South Africa is 'a global player' and 'wants to be seen as being concerned about conservation' – whales, she said, 'are charismatic'. She then explained that the 'Japanese Ambassador and a heavy delegation came to my office' in the period 'pre-October 2004 – two days before the CITES CoP'. They were, she said, 'wanting support' and 'asking this *very* question', saying that they 'support us on ivory' and asking 'why can't you support us, if you're supporting the principle of sustainable use?' (Mbengashe, pers. comm., 2006). That the Japanese Ambassador approached Mbengashe in the lead-up to the 2004 CITES CoP, in order to point out a perceived contradiction in South Africa's stances on whaling and on ivory, confirms the importance of linking the two species, the two issues and the two treaties.

I asked Oosthuizen whether he had ever been approached by Japan on the issue of linkage between whales and elephants. He said that he had been, but that South Africa had 'made it very clear' at the IWC that it sees no inconsistency in its voting patterns concerning elephants and whales, and that he believed that the Japanese now recognise and accept South Africa's position. However, he added,

> they will of course still try to link whales and elephants in a debate if they feel it will enhance their position, but that is just part of international debate; even if they understand our position fully, and are in agreement with it.

Asked to explain what had happened when Japan had approached him on the specific issue, Oosthuizen told me that:

> a few years ago, before I was commissioner, when I was still the scientific adviser, official Japanese representatives used to meet with the South African commissioner asking for explanations on our voting patterns and on why we did not support the Japanese position.

Since then, he said, 'they still ask me occasionally during the IWC meetings to explain our position on specific points, but I keep on giving them the same response'; and added that 'every now and then they will try to put pressure on us, for example, saying "listen you're not consistent with your policy on elephants", but that is the case not only with Japan but with many other countries as well' (Oosthuizen, pers. comm., 2006). Kleinschmidt too told me that he had been approached on this. He said that embassy officials from Japan had approached him in South Africa also, as had officials from the Norwegian embassy (Kleinschmidt, pers. comm., 2006).

South Africa has arguably chosen a stance which avoids a contradiction between its policies on elephants and whales – insisting that the two species raise different issues. Whether this stance is correct or not is debatable. It might be possible to keep the two species, the two issues, largely separate, given the different fora in which they are currently dealt with, and certainly this is what South Africa is currently attempting to do. However, the separation is not easily maintained as there are issues common to both species, and overlaps between them, which will continually recur, and which will always make it difficult to keep a rigid distinction.

On linkage, I put it to van Schalkwyk that South Africa's environmental policy generally is that of sustainable use of natural resources, but that it might be argued that there is an apparent contradiction between South Africa's general environmental policy, in particular on the ivory trade, and South Africa's consistent opposition to commercial whaling. He expressed his view, perhaps with some understatement, that 'the situation at the IWC is a bit unique'. Explaining, he said that 'we've always expressed our support for the principle and the policy, the approach, of sustainable use; but what happened here is it became so polarised that it developed into a pro- and anti-whaling faction'. Significantly, though, he then said that:

> we've also always made it clear to the so-called anti-whaling group, or the like-minded group, of which we are part, that we are a sustainable use country and as a matter of fact in future we believe that a possible compromise and way to take the IWC forward will have to include a compromise on the issue of sustainable use.
>
> (van Schalkwyk, pers. comm., 2007)

This may well be the crux of the matter for South Africa. With a completed RMS for the RMP, or even with an RMS that was *clearly* complete but on which it was *manifestly* obvious that only political stalling stood in the way of its adoption, South Africa might find itself under increasing pressure to support sustainable usage of whales – even where this meant consumptive use.

8 Sovereignty and environmental damage

8.1 Sovereignty

It can be argued, strongly I suggest, that there are certain issues which must transcend national sovereignty, being of extreme importance to the global community. The protection of biodiversity in rainforests, overfishing of the oceans and the future of species such as the elephant and the rhinoceros must be leading examples of these. Strangely, however, a state might be adamant in its stance on a particular international issue in order to gain certain short-term advantages; even if it does not really believe in its own professed stance.

It seems that at the opening session of the IWC in Washington DC, on 20 November 1946, the US Secretary of State, Dean Acheson, said that 'whales belong to no single nation nor to any group of nations but rather they are the wards of the entire world' (Sand, 2008: 58). Of course, in context this was probably intended largely in a utilisation sense – despite the apparently protective word 'wards'. At IWC 25, 1973, the US commissioner said that 'whales come under no nation's exclusive national jurisdiction, and as such have to be an international trust in whose disposition all nations should have a voice'. He then added that he did not think that 'the world any longer tolerates the view that only the nation or the industry which is in a position to exploit a marine resource should have full and final responsibility for it' (IWC 25, Verbatim Record).

Similar sentiments can be found in the opening statement to IWC 33, 1980, by the commissioner for St Lucia, who stated that 'we consider the oceans and all their flora, fauna and mineral resources as mankind's common inheritance', and that 'perhaps the time is also opportune for St Lucia to propose that the IWC become an affiliate of the United Nations under its Charter, and that discussion on the Law of the Sea and the delineation of the sea boundaries make reference to the whale as the common inheritance of mankind' (IWC 33, Opening Statements).

Nollkaemper suggests that 'with a seemingly schizophrenic persistence, states simultaneously pursue policies to protect forests, animals, and ecosystems located in the territory of other countries on the one hand' and, on the other, to 'support a legal system that protects states' sovereign rights to determine for themselves whether or not to protect such values' (1996: 237). It has, though, been suggested that even were it ever true that only the view, or 'writ' of a nation-state could run within its borders, that view can hardly be true today in current international life. This is because it has been demonstrated clearly, by the process of globalisation, per Friedheim, that states have great trouble controlling transnational business, pollution and communications and ideas within their borders (Friedheim, 2001a: 323).

It can be asked, in the face of globalisation, how true the sovereignty idea can really be in the twenty-first entury. Vogler, however, points out that 'despite various assaults on their competence nation states remain legally pre-eminent in the world system and it is therefore unsurprising that they should be the parties to global commons agreements'. The concern that can be expressed, however, is that 'they cannot be expected to rise above their own short-term national and electoral concerns' and 'ecology becomes the plaything of *realpolitik*'. He then quotes Sachs as saying, in 1993, that 'the rhetoric, which ornaments conferences and conventions, ritually calls for a new global ethic but the reality at the negotiating table suggests a different logic'; and there, 'for the most part, one sees diplomats engaged in a familiar game of accumulating advantages for their countries, eager to outmanoeuvre their opponents, shrewdly tailoring environmental concerns dictated by their country's economic position' (Vogler, 2000: 215–6).

Natural resources can sometimes stretch the boundaries of sovereignty. One might think that the elephant would, in all cases, be a national issue – since each individual elephant must be always within the boundaries of a state. Kreuter and Simmons, however, say that while the legal title to elephants, 'as with most other wildlife, is generally vested with a designated government agency' inside a country, 'given their fugitive nature, large home ranges, and widespread weak law enforcement, elephants have, however, effectively become an open-access resource in much of Africa' (1994: 39).

Bonner argues that the world is perhaps wrong to see elephants as being anything other than national property, writing that 'if private ownership is one extreme, the other is the assumption in the West that the wild animals of Africa, like many other wonders of the world, belong to the world'. 'Elephants and rhino, lions and leopards', he says, 'are not like whales living in international waters' as 'for the most part, these land animals live within the borders of a specific country'. Why, he asks, do these natural resources not therefore 'belong to that country, just as copper, coal and oil belong to a country?'. Do we, he continues, 'have more right to tell the Zimbabweans what to do with their elephants than we do to tell Saudi Arabia what to do with its oil?'. By way of an answer, he suggests that:

> if the wildlife of Africa is a priceless world heritage, if it belongs to all of us, and not just to the present generations, but to our children's children, then the world has an obligation to pay for its preservation

and that this is an obligation which has not been met – with it being 'too easy to impose bans – and make the Africans pay' (Bonner, 1993: 286).

This raises the question, of course, of the value that can – and perhaps should – be placed on symbols. Bonner asks the question of what would happen if

'Zimbabwe decided to kill elephants not for scientific reasons but simply because people were complaining that there were too many elephants?'. The issue, he says, 'is not just numbers, but sovereignty' and whether the international community has 'the right to tell a country what to do about its wildlife' (Bonner, 1994: 62). Along the same lines, Kreuter and Simmons blame the West, particularly the US, for not understanding the nature of sovereignty over natural resources. They suggest that increased antipathy toward the commercial use of wildlife, 'together with America's increasing involvement in international policy formulation since 1945', have created 'a national psyche which feels justified in imposing its non-consumptive values for wildlife on other sovereign countries' (Kreuter and Simmons, 1994: 42). According to Burns and Wandesforde-Smith, at IWC 53 in 2001 'several developing country parties also accused NGOs of seeking to undermine their sovereignty through threats of boycotts if they supported whaling' (2002: 200). So, can NGOs protesting against whaling be seen as threats to sovereignty, and the protection thereof? Certainly the object of such NGOs would have been the protection of cetaceans rather than a deliberate undermining of sovereignty – the effect, however, might be the same in the eyes of the affected developing countries.

It has been argued that it is only within the artificial parameters of the IWC that states accept – albeit reluctantly – such intrusions on their sovereignty. Friedheim contends that 'the principal difference between the IWC and many other international organizations is the parliamentary nature of the IWC's predominant decision-making system'. In the strictest sense, per Friedheim, 'one might characterize the IWC as a legislative rather than a negotiating system'. The underlying problem, he argues, concerns state sovereignty with majority representatives in the IWC having either overlooked the underlying sovereignty problems or having a 'hidden agenda' and using the IWC 'as a venue for attacking sovereignty'. For some stakeholders, he says, saving whales is the objective, while 'for others reducing sovereignty is what they are after' – and for still others 'these ideas reinforce each other' (Friedheim, 2001c: 207). It must be asked for which 'stakeholders' is reducing sovereignty the objective? Friedheim does not illustrate his point. The idea is again one of symbolism, of course, with the IWC as a battleground for shaping the world's natural resource use constructs.

Clover asks, 'who owns the sea?'. Fishermen, he says, 'like to think it is them . . .'. The real answer to who owns the sea is, he concludes however, 'everyone and no one', since 'if there is an owner of a common resource in a democracy, it is the people' (Clover, 2004: 220). This argument is reminiscent of national law of property, almost certainly true of any country, where ownership cannot be complete – no landowner can have absolute rights to use his or her property as he or she desires, no matter how absolute that landowner considers his or her ownership to be. The problem is that enforcement occurs only through a variety of uncertain measures.

Whales and elephants are, in one important sense, obviously different as whales are often not within national borders. Arguably, however, this does not make them 'free for the taking'. Gillespie suggests that 'cetaceans belong to no nation or region', but that 'this does not make them common property, free for the taking'. Rather, he says, they ought to be seen as being 'the property of the international community to whom the IWC, with its mandate to recognise the interests of the nations of the world in safeguarding for future generations, is responsible'. With cetaceans, he concludes, the 'primary guardian is the international community as manifested within the IWC' (Gillespie, 2002: 39).

Kalland argues that a 'strategy' of whale protectionists involves 'claiming that whales are not a free good owned by nobody until spotted, chased or captured', and that, in furtherance of this strategy,

> attempts are made to turn whales from *res nullius* (an open access resource) into *res communis* (a communal resource), which means that nations without previous interests in whales, Switzerland for example, come to share property rights and management responsibilities with the others.

The claim, he says, that whales are everybody's property is based on the notion that certain valued resources should be proclaimed 'the common heritage of mankind' (Kalland, 1994: 178).

The future of the IWC, Friedheim suggests, 'is an important test case of the world's ability to develop rules and rights based on shared norms in an issue area', and 'a failure in the area of whales and whaling might have a ripple effect'. What is at stake, he suggests, is not only whether aggrieved states will remain members, but whether the IWC can establish rules that will be obeyed, develop effective compliance mechanisms and whether proper leadership will emerge so that the IWC can escape its present state of 'stasis' (Friedheim, 2001b: 9). Certainly it is difficult to argue that the international community has the right within the borders of states to tell those states how to manage their natural resources. However, in respect of trade the issue might be different – especially where a treaty has been agreed to, and where the external community is affected.

The issue of sovereignty can be a serious point of contention. At IWC 54, in 2002, in a written statement, Antigua and Barbuda referred to the issue of whether Iceland should be allowed to adhere with a reservation to the moratorium on commercial whaling. In its statement, Antigua and Barbuda said that there are matters which:

> ought to be handled with extreme care. Countries are understandably jealous of their treaty rights under international law. If we abuse our own procedures, if we stand international law on its head, then we do nothing but bring our own organisation closer to the day when its credibility will be shattered and chaos will reign.
>
> (IWC 54, Statements)

Ultimately, whales and elephants stand on different footings, so to speak in the whale's case, in the sense that the former belong to an 'open access regime' and the latter within national sovereignty. Nevertheless, there is a cogent argument to be made that, despite this obvious difference, their destinies are mingled and many of the same sovereignty issues arise in respect of both. In a globalising world and where the primary end-users of elephant products are not located in range states, the sovereignty argument is far more complex than it initially appears. It is strongly arguable that the international community has rights and interests in controlling trade – and corresponding duties, and the 193 states party to the CBD have made international commitments to protect their own biodiversity.

8.2 The 'tragedy of the commons'

The world's understanding of, and attitude toward, the oceans has changed much over the centuries. Huig de Groot (Grotius) in 1609 described the oceans as being:

> that expanse of water which antiquity describes as the immense, the infinite, bounded only by the heavens, parent of all things; the ocean which the ancients believed was perpetually supplied with water not only by fountains, rivers, and seas, but by the clouds, and by the very stars of heaven themselves; the ocean which, although surrounding this earth, the home of the human race, with the ebb and flow of its tides, can be neither seized nor inclosed; nay, which rather possesses the earth than is by it possessed.
>
> (Grotius, 1609; in D'Amato and Hargrove, 1974: 1)

Today, a more realistic view of the ocean system 'is almost diametrically opposed to that of Grotius', with the oceans being 'very finite indeed' and constituting 'a complex and delicate ecosystem facing injury from many sources' (D'Amato and Hargrove, 1974: 1). As I suggested a few pages back, this needs to be seen in the same way that the Roman Law-style concept of absolute ownership is seen today. Although it is recognised that absolute rights in land are unacceptable and simply not congruent with social obligations, neighbours' interests, general national and international interests and environmental considerations, we still labour today under the influence of the concept.

Hardin's account of the 'tragedy of the commons' was that in the common grazing grounds of seventeenth-century England, each farmer bordering on the commons found it in his interest to add to his livestock; with the problem being that where each farmer bordering on the commons did the same the result was an overgrazed and destroyed commons, and economic ruin of all farmers (Hardin, 1968). The theory may be applied to the oceans instead of to the land; D'Amato and Hargrove writing that the ocean is a 'commons' in Hardin's sense. The point, in the authors' words, is that 'instead of individual farmers bordering on it, similarly acquisitive, self-interested, and one fears myopic, nations control the land masses bordered by the oceans' (D'Amato and Hargrove, 1974: 3–5). M'Gonigle explains, of Hardin's theory, that the tragedy of the commons occurs when ownership of the resource is held in common; a large number of users have independent rights to the use of the resource; no one user can control the activities of other users, or, conversely, voluntary agreement or willing consent of every user is required in joint action involving the community of users; and when total use or demand upon the resource exceeds the supply (M'Gonigle, 1980: 123–4). As Hardin himself wrote: 'ruin is the destination toward which all men rush, each pursuing his own best interest in a society that believes in the freedom of the commons. Freedom in a commons brings ruin to all' (Hardin, 1968: 1244). According to Vogler, 'following the logic of Hardin and other rational choice theorists, actors (who may be well aware of the extent of their long-term interdependence) will still behave towards common resources in a perversely individualistic and short-term way' (2000: 10–12).

The theory has been criticised. Hardin, Vogler explains,

> correctly identified the overgrazing problem associated with an open access *res nullius*, but the crucial point is that he does not recognize the possibility of a viable alternative to privatization or coercive management in the form of communal CPR institutions or the collective provision of public goods.
>
> (2000: 13)

Vogler does then offer a solution to the 'tragedy', writing that such solution is 'provided by the state', but that, strictly speaking, 'such alternatives are unavailable for the global commons'. Such areas, he explains, *are* global commons because 'they do not fall under the jurisdiction of any state and the defining characteristic of the international as opposed to domestic systems is the absence of a super-ordinate authority or world government' (Vogler, 2000: 15–6). If there is to be a solution to the 'tragedy' it might, it seems to me, be that it will one day be found to have had its beginning in the greater involvement of people in international law-making, and their greater influence on cooperative systems – vide the rise of influence of NGOs.

In any discussion of regulation of the utilisation of natural resources, it is important to have an awareness of the problem of the 'tragedy of the commons'. Arguably, all wildlife-related international MEAs and regimes have been put in place to deal with this problem.

8.3 The tragedy of overfishing

As early as 1882 it was recognised that the world could have a problem in respect of overfishing. In that year, Belgium, Denmark, France, Germany and the UK signed the North Sea Fisheries Convention and agreed to mutual rights of visit, search and arrest of the treaty parties' public vessels. This agreement is staggering when looking at the lack of protection given to the oceans in the early twenty-first century; after more than 130 years little has been done to offer substantive protection to marine species. Concern in fact began earlier even than the 1880s. In the UK, a Royal Commission on Sea Fisheries was held in 1866. This Commission, strangely, reported that 'fears of over-exploitation were unfounded' and recommended that 'existing laws regulating fishing grounds and closed seasons should be removed' (Sims and Southward, 2006: 660). Even more strangely, there are analysts who take a similar line today.

Before the Second World War, concern about overfishing could be seen in the 1937 International Convention for the Regulation of Meshes of Fishing Nets and the Size Limits of Fish. Although the Convention never entered into force, fisheries did receive a temporary reprieve during the six years of the war when there was very little fishing activity worldwide. After the Second World War, although a new fishing treaty was negotiated (in 1946) and entered into force (in 1953), fishing intensified dramatically worldwide.

It is incredible to me that, given that there was international concern expressed in 1866 and 1882, and again in 1937 and 1946, in relation to fish stocks, there is not now, in the twenty-first century, widespread – global – panic. In other words, if there was sufficient concern in the 1880s to take what was, for the time, the extremely radical step of holding a conference which produced a Convention, it is, to me, almost beyond belief that there is not in the twenty-first century great alarm when little or nothing has been done to solve the problem despite our environmental awareness and knowledge having increased exponentially many times. I can only guess that because the disaster is of a global scale it has not been recognised by policy makers and people generally, both of whom tend to view things regionally or locally.

It is also a problem which has many causes, which makes it easy for those guilty of one but not another to find a scapegoat. The causes include acidification and eutrophication of the oceans, bycatch and consequent wastage of fish species not targeted (including the damage done to sea-bird populations), climate change (which may have numerous side-effects little thought of, such as ocean thermohaline circulation and continental shelf 'flushing' and nutrient recycling), the creation of maritime dead zones, fragmentation of habitats, habitat destruction (especially of coral reefs and to sea beds from bottom-trawling), increased human coastal development, overfishing, pollution (in all its forms from raw sewage to heavy metals to acoustic pollution), sea temperature changes and the spread of alien invasive species. It is not any one of these in isolation, but their potential cumulative impacts that provide the most frightening possible consequences.

There are, of course, those who continue to deny that there is a problem, or at least that it is serious. The economist Lomborg writes that 'the global fish catch has in the 1990s not increased as much as earlier', and that 'this is primarily because the world's fishing fleets have a general tendency to over-fish particular stocks' with it being 'estimated that about 35 percent of the fish catch is obtained from stocks showing declining yields' with this being 'due to a mechanism with which we are all familiar: everyone's responsibility is nobody's responsibility'. The oceans, Lomborg says, 'could produce 100 million tonnes of fish a year' which we can harvest 'for free' (in the sense that we do not have to feed them). Right now, however, he states, 'we only catch about 90 million tons, the missing 10 million tonnes being the price we pay for over-fishing the seas'. So far so good, but he then says that 'we would love to be able to get hold of that extra 10 million tonnes, but this in no way decides the outcome of food availability', since:

> even if the world's fisheries could be perfectly orchestrated such that we were able permanently to catch the extra 10 million tonnes, this addition would be the equivalent to the increase in the rest of the world's agricultural production over the next 19 days, measured in calories.

Thus, he concludes, '*not* catching the extra 10 million tonnes is inefficient but in effect equivalent to just putting the world food development back a bit less than three weeks' (Lomborg, 1998: 107–8). This view does seem appallingly naïve. It appears merely to be a simplistic equation which ignores any extrinsic considerations, such as damage to biological diversity, effects on ecosystems, the numerous dangers of fish farms, the ability of the environment to support fish and so forth.

Lomborg aside, even proponents of increased consumption of natural resources (this might be objected to as an unfair characterisation, but what is meant is that such proponents argue against preservation and in favour of increased consumption – albeit *better managed* increased consumption) generally agree that overfishing is a major problem. According to Lapointe, 'over the past two centuries, the growing sophistication of fisheries technology rapidly increased harvesting capacity and processing efficiency', and 'with the technological advances in capture techniques, as well as the ability to process great quantities of seafood on board catcher-processor ships at sea, the world's commercially desirable marine species took a tremendous beating' (2003: 14). As Ellis puts it, 'a million vessels now fish the world's oceans, twice as many as there were twenty-five years ago'. He then questions whether there are twice as many fish as before (Ellis, 2003: 13).

The serving of selfish, short-term interests is everywhere. Clover writes, for instance, that 'if Europe depended on its own seas for food, draconian solutions to save the fish stocks and return them to historical levels might have been imposed by now', but 'we live in a global market so Europe simply exports its demand for fish, with all the destructive power this has had in its own waters, to other oceans of the world' (2004: 17). One might comment cynically, however, that this is exactly what Europe *is* doing – by exporting its demand it is able to conserve the fish in its own waters.

Of course, those who argue that there are 'still fish in the sea' do have a point. Probably the overall 'tonnage of life in the sea – what scientists call the biomass' has not declined much, but what has changed 'is the relative balance between species', so that the greatest weight of living creatures is no longer made up of 'long-lived animals, such as the skate, cod or oysters, but small, short-lived creatures, such as prawns, langoustines, Dublin Bay prawns or scampi, sand eels, starfish, jellyfish, worms and plankton' (Clover, 2004: 56). Perhaps, then, humans will be able to stave off a collapse in food security by turning their attention to the eating of animals which they do not currently 'fancy' – the real danger, however, might be that it is not simply the appearance of the oceans which will be altered, but their long-term health which will be irremediably ruined.

A major problem of fisheries management has traditionally been inadequate understanding of the correct, or at least the best-known, approach to take to management. The management solution most commonly offered is that of maximum sustainable yield (MSY). Dahmani explains the concept of MSY as referring to 'a constant amount of catch produced by a fish-stock at a given effort of fishing'. 'As far as the fishermen are concerned, therefore', he explains, 'MSY necessarily involves the restriction and control of their fishing effort at a given level, beyond which the fish-stock will be over-exploited' (Dahmani, 1987: 48). MSY, per Pauly and Maclean, is 'predicated on the idea that, up to a point, catching fish actually increases the amount that can be caught' (2003: 65).

In what state are global fisheries in the twenty-first century, truly? I have an impression that they might be in even more trouble than is realised. Historically, a peculiarity of collapsed fisheries seems to be that they thrive just before collapsing.[1] I have an idea that such a compensatory boom might even be what we are in at present – but that because it is happening on a global scale we do not realise it. The damage (per researchers Myers and Worm) was done in the 1950s and 1960s (Sample, 2003); arguably, today, we are simply mopping up the last few (commercially viable) fish. Obviously, this impression cannot be a scientific one – I am not a scientist. My view is based on the innumerable accounts of falling catches worldwide, which far outweigh those of massive harvests; that in talking to fishermen one always gleans much the same story (that fishing today is 'not what it used to be'); and on the sheer logical improbability of the current global fishing effort (with radar, sonar, satellite mapping, global positioning systems and electronic fish finding systems as aids; the oceans being virtually fully mapped; weather and seasonal patterns being largely understood and predictable; and the ever-increasing demand) being sustainable.

What prospects are there of saving global fisheries? Clover comments that 'seas managed as commons that no one is responsible for are the most likely to fail' whereas 'seas controlled by people who know who owns what and have legal responsibilities and assets are the most likely to succeed'. Not all commons fail, he explains, 'but giving rights and responsibilities to individuals or small groups of named people generally improves the fishery' (Clover, 2004: 269). The reality, however, is that the world is seeing increased encroachment onto areas within territorial control – not less. The majority of the world's fisheries are to be found within states' EEZs, rather than on the high seas, and the chances of a significant number of states altering the present legal regime, or suddenly gaining the ability effectively to enforce control of territorial waters, and of doing this while there is still time to save the fisheries, are slim.

1 Clover writes, for example, of 1997 that 'nature had just delivered one last, miraculously large year-class of cod in the North Sea' (2004: 150). Clarke writes, of the collapse of the pilchard (sardine) fishing grounds off Namibia (then South West Africa), that the take crashed from 1.4 million tonnes in 1968 to less than 10,000 tonnes in 1980 – and that it has never recovered. Of interest is that after the catch had slumped in 1971 to 646,000 tonnes, it rose in 1973 to 706,000 tonnes – but this 'final upward surge' happened because 'under pressure nature tended to compensate by rapid breeding' (2002: 169–72).

The aim of global players seems to be to restore fish stocks, rather than significantly to change patterns of consumption. Clover points out, though, that nothing says that the 'aim' of restoring stocks will work, just as the aim of 'not crashing them' has not worked 'for the past 15 years under European management, as stocks have sunk lower and lower' (Clover, 2004: 275–6). By way of illustration of just how outmoded fisheries regulations can be, consider the fact that in the European Union, at least until early 2013, it was illegal to land undersized or endangered fish, and so most such caught are discarded (dead or dying) as bycatch – by contrast, in Norwegian waters it has long been illegal to discard such bycatch.

Generally, instead of cutting back on fishing, the fishing industry appears to be seeking new ways to exploit previously 'underexploited' fisheries and species, and there is also a tendency to see fish farms as a solution, despite the various problems associated with these hatcheries. The argument has been made that although one must be concerned about this, and desperately wish to repair the damage done, this is now impossible and one should accept that the seas are no longer capable of being remedied. It is time, therefore, for an ecological shift in man's relationship with fisheries. Marra, for instance, has argued that fishing in the ocean is 'no longer sustainable' and that 'worldwide, we have failed to manage the ocean's fisheries – in a few decades, there may be no fisheries left to manage', which raises the question of what ought to be done? He suggests large scale domestication of the ocean. Marra in fact argues that this is happening already, arguing as Lomborg did that it is aquaculture that has been 'entirely responsible for the increase in world fish harvests that has occurred in the past 18 years' and 'marine aquaculture, or mariculture, is also growing, and given eventual limits to space on land, moves towards expanded mariculture, will accelerate in the years to come'. Many problems have been cited with fish farming, and Marra concedes that careful consideration is needed of how this domestication should happen 'to avoid many of the pitfalls cited by environmentalists and scientists'. 'But if we recognize', he concludes, 'that domestication of the ocean is starting to happen, we can craft a research agenda to mitigate the problems and maintain both economic and ecological sustainability' (Marra, 2005: 175–6).

Fishing remains an area in which there has been little movement in international law. The UNCLOS of 1982 (which entered into force in 1994) settled states' EEZs at 200 nautical miles. This provides an important protective measure, as the majority of marine living resources are to be found within coastal waters rather than on the high seas. The high seas remain, however, an open-access commons. In the Plan of Implementation, parties to the World Summit on Sustainable Development (WSSD) of 2002 committed themselves, inter alia, to the following:

> to achieve sustainable fisheries, the following actions are required at all levels: (a) Maintain or restore stocks to levels that can produce the maximum sustainable yield with the aim of achieving these goals for depleted stocks on an urgent basis and where possible not later than 2015.
>
> (WSSD Plan of Implementation, 2002)

To me this halfhearted commitment merely staved off for a while longer acknowledgment that drastic steps need to be taken urgently to prevent complete and utter, rather than just devastating, collapse.[2] Overfishing of the world's oceans is today a greater problem than it has ever been. The efforts of the drafters of conventions aimed at the protection of fisheries have not kept pace with the technological expertise, or the energy and drive, of fishermen; and nor have the various marine pollution conventions proved effective in combating pollution of the seas, either by way of dramatic oil spills or the continual hazard of daily operational discharge from the countless vessels plying the seas. The biggest danger must surely be that a global crash in fisheries will occur so rapidly that there will simply not be time for the world to respond – the crash could even be from one year to the next.

As much as the parties to the ICRW wish, for their various reasons, to keep what is arguably the world's most dysfunctional MEA to itself, it is ultimately impossible to divorce consideration of whaling from what, in my view, is one of the greatest tragedies of all time – the selfish, short-sighted and almost complete destruction of the oceans as a set of properly functioning ecosystems. The imminent catastrophe of the worldwide collapse of the oceans' ecosystems bears directly on the whaling issue for many reasons.

The sense has grown on me over the years that in many ways the battle over how to utilise and how to protect whales and elephants is representative of the battle over how to utilise and protect *all* natural resources. In particular, it sometimes seems to me that the focus on whaling – with its various thrusts, feints, withdrawals, threats and concessions; and with all of the financial resources and emotional energy that are put into the battle – may really be just a huge red herring. For Japan, perhaps the whaling issue is a 'line drawn in the sand' (or the water!) in that the country may feel that if it compromises on whaling then it will immediately find tuna becoming the next battle, and then sharks, and so on. Or perhaps the country feels that so long as it keeps Western environmental NGOs pouring their energies and resources into the whaling issue, it will be left alone to plunder other marine resources (strangely, I find it hard to decide whether either of these views damages or assists my argument that whaling has important symbolic value for both sides). It is possible also that, as Kleinschmidt argued, Japan has accepted that the fish resources of the oceans are now too far gone to be saved and, looking ahead to a time when whales will be seen by all as a necessary food resource, wish to keep in a state of preparedness to exploit this resource.

2 What is perhaps worse is that in mid-2012, in the Rio+20 outcomes document 'The Future We Want', this commitment was not scrapped but renewed with the words: '[w]e commit to intensify our efforts to meet the 2015 target as agreed to in the Johannesburg Plan of Implementation to maintain or restore stocks to levels that can produce maximum sustainable yield on an urgent basis'.

9 Different palates

Oriental and Occidental

There are several extremely wide rifts, or gulfs, in understanding between different protagonists in the conservation debate. Likewise, on the international plane, there is an enormous gap in understandings between – put crudely – East and West.

Watson writes scathingly that 'it's all well and good to preach the save-the-whales message to 300 million North Americans: they aren't killing whales' (Watson, 1994: 24). According to Chadwick, 'in Bangkok, people can still order a feast with an Asiatic black bear or a sun bear as the centrepiece, bawling and screaming as it is slowly roasted alive over coals before the assembled guests', or a restaurant 'will cater one of those affairs, known throughout much of the Orient, at which the top of a live monkey's skull is removed and the primate watches the diners scoop its brains out to eat' (Chadwick, 1992: 364). Perhaps there is a difference between a society that will import endangered foods from other countries – to the detriment of the species concerned – for no reason other than to satisfy a taste for the exotic, and a society that kills the same endangered species simply for subsistence. However, the lines do sometimes become blurred. While Chadwick criticises an Oriental taste for food that involves unnecessary suffering, such criticism makes it difficult to answer the Japanese charge that Japanese whaling is for food, that no parts of a whale are wasted and that historically it has been the West that used no parts of whales for food and which has wasted the parts of whales that they did not intend to use. Different sensibilities abound, then. How to breach these gulfs in understandings, empathies and cultural practices may be the greatest conservation challenge of the twenty-first century.

Japan, at IWC 46 in 1994, argued that by the year 2050 it is estimated that the world's population will exceed ten billion. 'In order', suggested Japan's commissioner,

> to maintain our populations' food resources in this rate of increase, if we are to acquire all the food resources from the land area, we must cut down the forests and make more room for the agricultural and pastoral lands, and this inevitably leads to the environmental destruction of the global scale.

The commissioner then suggested that the world will eventually have to consider 'the poverty and shortage of nutrition and starvation suffered by the developing countries' population' (IWC 46, Verbatim Record). In its (written) opening statement at IWC 54, in 2002, Japan said that 'the whale to the Japanese in ancient times was a kind of fish that was thought to be brave and great', and that:

> the total protection of all whales irrespective of their stock status as promoted by some members of the IWC and some NGOs is contradictory to Japanese cultural values where whale meat is still eaten and where whales are still revered through religious ceremonies and festivals.
>
> (IWC 54, Opening Statements)

Komatsu and Misaki write that 'whales are a wonderful gift from the sea' and that 'they are among the most favourite of the marine food supplied to Japanese people and are traditionally an important food resource', and add that 'whaling today, and in the future, will be for food, not oil' (2003: 6–7, 21).

Komatsu and Misaki suggest that 'there is another factor that singles out Japan in the whaling issue, and it is a racial one', with Japanese believing that because they are 'non-white', racial discrimination is 'a hidden agenda of the anti-whaling powers' with the 'imposition of Western values on Japanese' being 'cultural imperialism' (2003: 23–5). Certainly, the West has created a situation where it can be charged with hypocrisy – due to its past rapacious and wasteful whaling practices, and its current support for aboriginal subsistence whaling. While the West cannot 'undo' its past whaling practices, and while it might argue that it has come to a new understanding of whales which the Japanese ought to adopt, what it can do, in practice, is to approach the matter with greater understanding of the Japanese position and ensure at the very least that its behaviour in the twenty-first century is beyond reproach.

The position becomes even more complicated when one looks further into the Japanese attitude towards utilisation of animals for food, and of whales in particular. Komatsu and Misaki write that Western people 'would rather not know where their food comes from', and that, in contrast, 'there is a customary Japanese prayer said before a meal, [*i*]*tadakimasu*' which 'is an expression of gratitude for the very life of the creature that has died for our survival'. All killing of animals, they say, 'is cruel, but there is no justice, no logic in condemning the killing of one animal by one people, while living off the killing of another', and the 'best that we can all do is to appreciate the life that has been surrendered, and not waste any part of it' (Komatsu and Misaki, 2003: 154, 155). Yasunaga writes that there are historical records of memorial services for whales being conducted; both after significant takes of whales and at the ends of whaling seasons. According to the writer, such services reminded people that 'people should not forget that they make their living by taking the lives of whales' and that 'there is no difference between the life of a small fish and a large whale' (Yasunaga, 2005: 81).

Iino and Goodman write that 'from a Japanese perspective, whales are valued marine resources that should continue to be used as a source of food, provided that the stocks are robust enough to support sustainable whaling' (2003: 7–8). Komatsu and Masaki argue that 'when one does something by tradition, one is unlikely to deeply question it', and that, with the eating of whale meat being 'ingrained in the Japanese tradition', the concept of 'cruel killing' is 'not really one that enters the Japanese mind'. Neither, they argue, 'is it one that enters the Western mind, when contemplating the beast that met its fate in a slaughterhouse to provide them with beef or lamb' (Komatsu and Misaki, 2003: 105–6).

It can reasonably be asked why the issue is of such importance to Japan. In the words of Iino and Goodman, 'many commentators question why the Japanese are so adamant about the right to whale when whaling does not contribute significantly to GDP and whale meat is no longer a significant part of the diet of most Japanese', explaining that the 'short answer is that it is a matter of principles' (2003: 8). This suggestion in fact enhances the point made above: that the things that are at stake are the symbolism of the whale, and the pride, stubbornness and self-interest of states. Friedheim suggests that 'to force Japan out of whaling seems to Japanese another measure to reduce their independence and create a sense of dependence on outside food sources' and that it is, therefore, 'little wonder they have been stubborn'. At the same time, however, Japan has – Friedheim argues – 'always bargained sincerely', giving as an example 'the fact that they did not vote strategically on American requests for an aboriginal quota' (Friedheim, 2001a: 328). Unfortunately, this suggestion of sincere bargaining ignores the events (which give the lie to it) surrounding the bowhead whaling controversy in the late 1970s, and, being published in 2001, precedes the events of the May 2002 blocking of the renewal of aboriginal bowhead whale quotas – at which point Japan most certainly *did* 'vote strategically'. This contention further ignores the possibility that Japan never did intend to abide by the 1982 moratorium to which it agreed to be bound – in the sense that the country may have agreed to the moratorium in the knowledge that it would legally be able to rely instead on scientific whaling, until the moratorium could be overturned.

According to Chadwick, conservationists have 'put a lot of effort into trying to analyze the Japanese view of the living world', asking questions such as: why is 'this nation so willing to keep accumulating ivory when to do so is to visit disaster upon elephants?'. Continuing, and attempting to provide answers, Chadwick writes that 'Japan is synonymous with an extremely refined appreciation of nature'; however, 'the operative term here is not nature but extremely refined, as epitomized by the stunted, root-starved, strapped, and carefully twisted little trees called bonsai'. Wild things and places, he continues, 'apart from the human sphere are something else again', and 'Japanese culture has not traditionally accorded them much sympathy or respect'. His argument is that:

> Japan is placing exceptional pressures on the biosphere just now not because of some rapacious streak in the national character, and not because its citizens are exceptionally greedy, but simply because the Japanese are humans and their country has an exceptional concentration of capital and technology at the moment.
>
> (Chadwick, 1992: 221–2, 222–3)

As illustration, however, of the gulf of understanding which exists, one can consider the views of Akio Morishima, whom I interviewed. Questioned on the role of Buddhism in the debate, Morishima answered that in Buddhism 'all animals are respected – fox, bear'. He then suggested that the Japanese 'use all parts of the whale' and that the whale is 'not very special'. Questioned as to the concept of nature in Japan, with me putting it to him that the bonsai tree represents an 'unnatural' view of nature, Morishima explained that 'Japanese gardens are unlike European gardens'. In his words:

> we don't cultivate gardens as Europeans do. We don't line trees. Put it in a natural way. Take nature as it is. Bonsai is nature's way. We don't see humans as having a special existence – humans are like dogs. We are part of nature. Traditional perspective. Born from it, come back to it. We might be a dog in future. If I live a good life, could have a good future life. Cycle. Even the mountain or stone could have a life. Don't have one god. We don't have an idea of nature conservation because we are part of nature. Harmonise with nature. Should not enclose selves and separate ourselves from animals. To survive, humans must harmonise with nature.

'Understand', he continued,

> why we have bonsai. Bonsai is *not* artificial, we don't like very artificial nature. Stay in small inn, from your room can see surrounding hills and it's a part of the nature. When look at the bonsai, it's just taking that scenery into your room, miniaturising all landscape into your life. British, French gardens, when I saw them for first time I wondered why they made them so artificial.
>
> (Morishima, pers. comm., 2004)

If the opposed camps in the 'preservation' versus 'conservation' war are ever to be brought together, there will need to be middle ground found as to the perceptions which each camp holds of the value of nature. At present, however, it seems unlikely that there will be agreement reached any time soon – so vast are some of the differences.

10 Biodiversity

10.1 The importance of biodiversity

From all sides of the debate, protagonists today depict conservation – in whatever sense they mean the term – as being the control, protection or management of interlinked species. No serious commentator would, today, appear to question that such interlinking of species is what makes biodiversity so difficult to understand and so complex a thing to protect or manage. Nor, today, is it seriously questioned that both protection and management are vital concerns.

'The real challenge for a global programme to protect biodiversity', write Schücking and Anderson, 'is to examine linkages'. A global programme, they say, 'is necessary precisely because environmental problems in the Third World (and elsewhere) cannot be solved in a national context', and because, 'given the global interlinkages of the biodiversity crisis, no single approach to conservation can be effective' (Schücking and Anderson, 1991: 34–7). Leite, *et al.* write that 'the environmental crisis has reached a new dimension' and 'has become a global and complex crisis, an undivided and interdependent part of a larger crisis'. They suggest that this larger crisis has been called a '*poly-crisis*', within which it is 'not possible to establish a hierarchy of concerns because there is not just one critical problem, but many interlinked critical problems causing the worldwide crisis' (Leite *et al.*, 2007: 471). Ultimately, this is my feeling too; that neither 'conservation' nor 'preservation' *in and of themselves* will be able to solve the global biodiversity crisis – but that where both are used, supporting and sustaining each other, there might be a possibility (albeit slight) of finding solutions.

There is certainly an argument to be made that at least certain areas of natural complexity ought to be preserved intact as far as possible – in other words, not managed, not used and left virtually pristine. Clover writes that Ballantyne contends that 'whether reserves would help fisheries is not the point'. The point, he apparently says, 'is that there should be areas of the sea where we see what nature is like in the absence of human transformation' with the fact being, according to Ballantyne, 'that choosing reserves because you want to protect this or that is a mistake', and the 'fundamental thing is to protect some of the marine environment' through the very simple idea of 'leaving parts of the sea alone' (Clover, 2004: 216–29). Chadwick points out that in reality, 'long before the brink is reached, and usually before a species' plight even starts to attract widespread concern, a substantial portion of the genes it carries will already have gone extinct'. He then argues that 'the metaphor of the canary in the coal mine may be terribly overworked' but suggests that we 'try thinking of elephants in that context', and asks 'if a trumpeting, five-ton canary cannot hold our attention, what can?' (Chadwick, 1992: 304, 470–1).

On the problem of preserving the biodiversity of elephant habitat, Chadwick writes that 'the problem is not just that elephant habitat is shrinking in terms of absolute size', but that it is 'being fragmented at the same time, shattered into discontinuous shards'. He goes on to say that this also has serious consequences for other species which live within elephant ranges, with a useful explanation for this being found in the ecological theory described as 'island biogeography'. The longer an island has been isolated, he advises, 'the less common features its flora and fauna will have with communities on the mainland' and such areas struggle to hold enough members of, especially larger, species needed to maintain a stable gene pool (Chadwick, 1992: 468–9). Parker essentially agrees with this observation, writing that island biogeographic theory is 'compelling' as national parks in Africa 'are all becoming islands to which the theory will apply'. No elephant population yet studied, says Parker, has been found to be stable or 'in balance' with its environment (Parker, 2004: 394–6).

The issue is extraordinarily complicated. Thomson tells us that the degree of 'controlled protection' needed by each species, or the degree of 'controlled use' which can be made of each species, may vary 'with considerable rapidity', and, consequently, 'the legislation giving force to management controls has to be elastic enough to accommodate change'. 'The decision *not* to interfere', argues Thomson, 'must still be recognized as being just as subjective and man-made as any other management decision' (Thomson, 1986: 43, 73).

Little more than 50 to 100 years ago, states parties to international conventions were classifying certain species as 'noxious' and recommending their extermination. It would be extremely dangerous for us now to claim that we understand how nature works. Almost certainly, in 100 years' time, the scientists of the future will be mocking our lack of insight. Despite this, in my view, we continue to fail to recognise the importance, complexity and fragility of biodiversity.

10.2 Whales, elephants and the 'ecosystem approach'

It has been suggested that the increasing abundance of the minke whale in the Antarctic Ocean is driving the blue whale from its diet of krill. This is actually not a very new argument. Writing in 1980, M'Gonigle suggests that at the 1977 IWC meeting it was argued that the minke whale's ecosystem had been 'turned upside-down by the depletion of competitor whale species' (1980: 151). According to Lapointe, more recently, blue whales are one of two great whale species 'that have not yet[1] rebounded sufficiently from over-exploitation'; as part-explanation for which he suggests competition for food with minke whales (2003: 80).

1 Lapointe is perhaps somewhat optimistic with his use of the word 'yet'.

Burton tells us that in one day 'a blue whale eats four meals comprising about four tonnes, which is about eight million krill, and a fin whale eats just under three tonnes'. This sounds a vast quantity, but Burton (writing in 1973) then goes on to explain that where in the past there were one million whales in the Antarctic each summer, now there are 'no more than a third of this number' with the original population consuming 'millions of tonnes of krill'. With the blue whales, right whales and humpbacks having been almost wiped out, and other species greatly depleted, he argues, there 'must be a vast uneaten surplus of krill, which, according to some estimates, could amount to something like 150 million tonnes per year'. What, he asks, is happening to this presumed surplus? (Burton, 1980/1973: 57–9). Of course, it might even be that whale populations in the past were far higher than is presently thought, and even estimated scientifically, to have been the case (*cnnonlinearchives*, 2003, quoting Palumbi and Roman).

Gambell agrees that there is no doubt that the removal of excessive numbers of whales has had a 'profound effect on ocean ecosystems', but he has a different take on it than has Burton. 'In the Antarctic', Gambell contends, 'many of the larger animals compete for krill as a source of food. As a consequence, the depletion of the stocks of the great whale species has had observable effects on the growth and reproduction of other species'. He concludes with the argument that we cannot with any current direct action 'redress the balance', and that all we can do now, for the time being, is 'watch and wait' to see if the great whales can recover their ecosystem places (Gambell, 2003: 82).

Some writers contend that whales are causing significant damage to ocean ecosystems, and are competing with fishermen. Komatsu and Misaki argue that the great amounts of fish that whales consume cause serious problems in human fisheries; the authors arguing that whales annually consume between 300 million and 500 million tonnes of fish, and that this is equivalent to three to five times the annual amount (approximately 100 million tonnes) taken by humans. 'On the other hand', the authors suggest,

> the Japanese fishing production has been reduced to a half of what it was 20 years ago. There is a strong case that over-protection of whales is causing detriment to the Japanese fishery. In greater perspective, this shows that unilateral protection of an animal at one level of the food web causes great damage to the ocean ecosystem.
>
> (Komatsu and Misaki, 2003: 7–8)

The same writers then claim that in the Antarctic the 'different whale populations are segregated and do not mix'. The area, they say, nearest to the South Pole 'is inhabited, first, by penguins, then seals, blue whales and minke whales' and that 'these different species compete over krill, thus forming the Antarctic biological ecosystem'. To the North of this, they argue, there are fin, humpback and sei whales and that 'it is nature's way that they should remain segregated' with this being 'the wisdom of nature – to segregate is to eliminate competition over food' (Komatsu and Misaki, 2003: 38).

I prefer an argument that 'peaceful creatures' like whales – the equivalent of the African bush's herbivores? – live side by side and eat differently as do impala, giraffe, kudu, wildebeest, zebra and other herbivores, on land. These species simply do not compete in ordinary, non-stressed, conditions – eating, as the impala, wildebeest and zebra do, the grass at different levels or types; and the giraffe, impala and kudu eating leaves at different levels or eating different types. Even in certain stressed conditions, such as at times of drought, the different species tend to cluster together – the advantages of communal living (especially enhanced vigilance) outweighing the disadvantages of competition for food.

At IWC 51, in 1999, Japan introduced a paper which estimated that total food consumption by cetaceans is three to six times the worldwide marine commercial fish catch (IWC 51, Chair's Report). If correct, this is an alarming statistic as the worldwide annual fish catch is probably currently in excess of 100 million tonnes. 'Alarming' it is, but it must be put into context. Whales excrete, die, feed other species and alter krill populations, and, importantly, there are many hundreds of thousands fewer whales than in the pre-whaling era. To give an idea of just how pitiful is the plight of the blue whale, consider that Hoyt suggests that in the Antarctic whalers killed some 360,000 in the first half of the twentieth century (Hoyt, 2005) – and that we now number their populations in the mere thousands, at best. At IWC 53, in 2001, Japan argued again that 'the slow recovery of blue whales', as evidenced 'by the Scientific Committee's SOWER (Southern Ocean Whale and Ecosystem Research) programme', indicates that competition for prey by minke whales shows the latter species continuing 'to prevail in the Antarctic' and that minke whales 'therefore need to be managed appropriately under the RMP/RMS' (IWC 51, Chair's Report).

Komatsu and Misaki argue that when blue whales were 'hunted down to depletion', the 'fast breeding minke whales took over the feeding ground', and that 'with increased availability of food, they rapidly' became dominant. 'It is believed', they claim,

> that this hampers the blue whales' recovery. It is the work of nature that the age at which minke whales, through better nutrition, reach sexual maturity has been lowered to six to eight years, resulting in a higher rate of breeding.

Their final assessment is that it might become 'necessary to cull the Antarctic minke whales to an extent where its dominance of the feeding habitat is reduced, allowing blue whales a greater opportunity for recovery' (Komatsu and Misaki, 2003: 38–9).

It might be questioned whether it would not make more sense for minke whales to be breeding more slowly as their numbers increase. Further, if there are many more krill – nature 'abhorring a vacuum' and there being less predators on them – then how can there be increased competition?[2] The debate shows the limits of

2 Butterworth commented on a draft of this text as follows:

> your logic is at variance with basic scientific principles here. If a population is to increase where it has been steady, say, either its birth rate must increase or its death rate decrease.

scientific knowledge at the moment – both arguments can be made, and either side could be right. Hence, perhaps, the argument can be made strongly that the precautionary principle should come into play.

Hoyt explains that:

> in recent years, those in favour of marine mammal, shark and other large predator culling, as well as whaling, have sought to use the language and ideas of ecosystem-based management to argue for so-called fisheries protection – killing predators in a misguided attempt to protect commercial fish stocks.
>
> (2005: 4)

Butterworth told me that the assessment was made originally to explain unusual age patterning in minke whales, based on analysis of ear plugs. I had put to him the claim of minke whales breeding at a lower age, and said that it made little apparent sense to me. The issue, he said, is 'still controversial', but the basic line is that there are two ways to discern the age of maturity in whales, the first being to catch the animals and examine the females' reproductive organs. Second, he explained, age of maturity:

> has been correlated to something in the ear plugs, which lay down like trees, you have layers each year. What happens is, they're initially widely spaced to become narrowly spaced and the reason for this is at a certain stage the skull stops growing – not much space for this ear plug to expand in as it goes up, and this is correlated with the age at which they become mature. The idea is that if there's more food available they grow faster so they reach this age earlier; and the question is, is this a major factor in the way the population controls itself?
>
> (Butterworth, pers. comm., 2006)

He then explained that there are various features that control population rate, and that this potential factor needed to be considered. He explained that when this data was first considered, a decrease in the age of maturity was apparent:

> from about 1950 to 1970, remember, no minke whales had been caught then, you could only pick this up because we were catching minke whales in the 70s and 80s, and you could tell, if you got a 40 year old minke in 1980, you knew it was born in 1940 and from the transition phase of the ear plugs when it became mature, at age seven or eight or ten, the picture that came out of this

> What is posited for the Antarctic is that decreasing the number of large baleen whales allowed their food source, krill, to increase and that minke whales (and other species) took advantage of this by increasing their birth rates and growing. It is then the greater numbers of these species which, through the further krill they now consume, inhibits the recovery of the species such as blue whales originally depleted by whaling.
>
> (Butterworth, pers. comm. 2008)

> was that in 1950, which is about as far back as you can take this reliably, they were becoming mature at age eleven; and by the 1970s they were becoming mature at age seven, and since then it's stabilised – and this is direct observation from the transition phase – it has gone up slightly recently, but stabilised again. In simple terms, it was at eleven, it came down to seven by 1970, it then went up a bit to eight by roughly 1980, and then it stabilised for the last 20 years.
>
> (Butterworth, pers. comm., 2006)

It seems odd to me that the minke whale's age of reaching maturity might have lowered because of the species' increased population. It is important as an example of the arguments raised by the pro-whaling camp, and of this camp's general mindset. However, in my (admittedly non-scientific) view, it would seem to have little biological credibility. My understanding is that in healthy populations, maturity ages do not decrease.[3] It is when populations are stressed – or, as with elephants, when there are insufficient adult animals – that maturity ages lessen.[4] That the argument is raised is, however, extremely interesting given that, as already mentioned, at CITES meetings from 1989 the Zimbabwean representative, Martin, apparently claimed (as an indication of stock health) that elephants in Zimbabwe were breeding at a younger age than had ever been recorded elsewhere (Thornton and Curry, 1992/1).

In one sense, what is most important about this issue is that elephants and whales are so similar – in respect of certain populations of both, it is claimed that they are becoming too numerous and affecting their environments negatively. In South Africa, in particular, there has been a debate raging since at least the early 1990s over whether there are 'too many elephants' or not. While it has often seemed to be virtually an article of faith amongst government-affiliated conservationists that it is necessary to cull elephants, particularly in the Kruger National Park (KNP), there is much scientific and other opposition to this. The argument for culling is that when elephants grow too numerous within an enclosed area, their impact on their ecosystem begins to become negative and that culling is necessary in order to protect endangered species. An example I was once given was of an elephant having destroyed a Lappet-faced vulture nest in the Hluhluwe-iMfolozi Game Reserve in KwaZulu-Natal province. While this is a completely natural act for an elephant, the

3 Butterworth commented on a draft of this text as follows:

> you are at variance with much scientific evidence here. Decreased age at maturity is one of the mechanisms populations use when reduced to get birth above death rates so that they can regain original abundances. Dropping the age at maturity gives an effective increase in average birth rate in the population, and lower age at maturity is an expected response to the greater food availability at lower numbers which allows the animals left to grow faster individually and hence mature earlier.
>
> (Butterworth, pers. comm., 2008)

4 In efforts since 1995 to translocate elephants in South Africa, young bulls translocated to areas where there are no adult bulls appear to reach puberty sooner. This may evince itself more in behavioural problems than in physical aspects, but is a certain trend.

problem was that the nest was one of only ten nests then known in all game reserves in the entire province. Public pressure led to a moratorium being put on elephant culling in 1995, and no elephants have been culled since then. Instead, various contraception and translocation programmes have been tried in order to relieve numeric pressure. The moratorium has also been essentially replaced by a set of 'Norms and Standards for Elephant Management' in terms of which culling is permissible as a management option, but is a 'last resort'. The jury remains out on whether elephants are indeed damaging to biodiversity when they become too numerous, but my impression is that the protagonists who argue that populations will eventually be self-regulatory are slowly winning the argument. After all, culling was used for decades to control numerous other species in the KNP, but most of those (for instance hippopotamus and wildebeest) are now considered to have self-regulating populations. I think that eventually the same will be found to be true of elephants.

There are more reasons why it is crucial to be cautious with elephant management. Working out how to value nature, it can be argued, is crucial to conservation, in whatever sense the word is used. It has been suggested that 'unlike most other species in the marine habitat', whales are now accorded a 'wildlife status' which recognises 'their intrinsic value as species for their own sake' (Rose and Paleokrassis, 1996: 148). However, this seems to me to be far too simplistic and potentially even dangerous. Value must be accorded to all ecosystem components. The African elephant is a potent symbol of, and for, conservation efforts worldwide. However, the true value of wild animal species must be considered not as abstract symbols, but as essential components of their ecosystems. The elephant has even been described as a super-keystone species. It is truly the landscape architect of the African bush and many physical and biological aspects of its habitat are dependent on its continued presence. The fight to save the elephant can be seen, therefore, as a fight to save the elephant's environment too. The same can be said of whales.

In respect of whales, one might reasonably ask whether it is right – either ethically or in scientific legitimacy – to compare human use to whale use. Humans, after all, can hardly be seen as naturally being systematic predators of fish (in the same way that chimpanzees are not habitual fishers) and whales being creatures that fit into marine ecosystems because of many millions of years of evolution and natural selection; whales also having no choice, unlike humans, but to eat fish. Yet the argument is being made.

Lapointe argues that 'like over-harvesting, over-protecting is not conserving. To the contrary, both activities work against conservation' (2003: 11–2). This is, of course, a difficult argument to refute – that species should be extracted from ecosystems equably. However, what it is presupposing is that significant quantities will be extracted. The solution might well prove to be the extraction of less overall rather than increasing the take, albeit taking more equably. It has been suggested, however, that in the 1950s marine mammals in the ocean were eating seven times more fish than was being caught by humans – a ratio which fell to three times in the 1990s, because of increased human catches and falling numbers of marine mammals. When only the types of fish taken by human fisheries are considered, however, it seems that marine mammals and humans are currently on about par. This, suggest Pauly and Maclean, seems to indicate that 'some fishers and fisheries managers are using marine mammals as scapegoats for their own mistakes' (2003: 56–7).

Lapointe argues that when predator numbers 'grow greater than the predator-prey equilibrium formula, and prey species begin to disappear', as on land:

> so too in the oceans, equilibrium is lost when there are too many whales or sea lions, seals or sharks. The entire ocean ecosystem suffers. Ironically, it is the spirit of saving a single species that threatens the survival of many other species. (2003: 15)

It is questionable whether this is indeed correct. This, after all, is the thinking that lay behind the London Convention of 1900 and the Paris Convention of 1902, and of early wildlife managers who waged an extermination war against predators, but which thinking proved eventually to be flawed as a management tool. Lapointe's argument may well be similarly flawed. As with plains game on the African savannah, there is an intricate marine web.

Stone suggests that it is 'not unthinkable' that eventually the IWC might have to consider the effect of whales on human food security. The 'ugly, little discussed fact', he says, is that marine mammals, 'probably whales in particular, compete with humans for fish on an unclear but clearly huge scale'. He then asks whether 'an expanding humanity' might not in the mid-twenty-first-century become 'tempted to regard whales as ranchers do wolves – with public sentiment shifting accordingly'. And, he says, 'might we not someday be looking to the IWC to – flexibly – account for food needs by culling out excess competition?' (Stone, 2001: 271–2). My answer would be that by the time it comes to the stage described, the greater collapse of fisheries generally ought to have made it obvious that turning to eating whales would itself be only a short term solution. Much will have changed by the mid-twenty-first century.

Further, drawing from the figures cited, the question might be asked whether minke whales have become so numerous that they now outnumber the combined total of all great whales before most of them were killed. This does not seem likely.

Iino and Goodman suggest that various international documents, such as Agenda 21, 'call for the promotion of an ecosystem approach to the management of marine resources' (2003: 11–2). What this does imply is that the 'ecosystem approach' has been deliberately linked with the idea of removing human-cetacean competition. It is extremely ironic how use of the term 'ecosystem approach' has shifted between the protagonists. In the late 1970s, Scarff could write that 'the need for cetacean management based on ecological principles is not merely an idealistic refrain, but a matter of immediate and practical necessity' (1977: 412). By this he meant that preservation was needed – by way of a moratorium – and the ecological ground was that whales desperately needed to be given time to recover, if they could. Asked about this use of terms, Donovan felt that this debate is 'part of a much broader problem and that it had become a mantra to talk of an ecosystem approach', but that 'nobody really knows what it means, you get quite a lot of disagreement over what it means' (Donovan, pers. comm., 2007).

I put it to Kleinschmidt that it is odd, at the same time as we appear to be moving toward more holistic understandings, to see whaling as an almost entirely separate issue. He agreed that it was odd, and described it as another reason why the IWC is 'out of step' with 'the next chapter in marine management'. 'Our grasp', he said,

> of the ecosystem approach is so wanting, and therefore you've got to look at the ecosystem pyramid, and understand how it is that you need a top predator, you can't just say we'll remove something here, [because] then it's got a rippling effect on everything else.

He concluded that nobody knows what it is that causes animals, especially smaller animals, sometimes not to recover from disruptions to their populations – but that it appears to be human intervention 'that cuts such a strong slice out of this pyramid'. I think, he added, 'that it's in that context that in the end whales need to be seen, and I don't think we've got a clue what their role is' (Kleinschmidt, pers. comm., 2006).[5]

10.3 The CBD

As it becomes apparent that the ICRW and CITES may not be able to offer any greater protection to endangered species, or enable harmony between their parties, than they do at present, and as their weaknesses become more exposed, so newer treaties must be looked to. The 1992 Convention on Biological Diversity (CBD) is perhaps the best example of a new approach to environmental treaties. The preamble to the CBD affirms that states have 'sovereign rights over their own biological resources'. However, it also affirms that the conservation of biological diversity is 'a common concern of humankind' and that states are 'responsible for conserving their biological diversity and for using their biological resources in a sustainable manner'.

5 Kleinschmidt clarified what he meant by explaining that what he had wished to convey with his comments was that the environmental concept of sustainability is often misused. By way of example, he explained that if there were once 1 million whales then the sustainability model might say that we can kill 100,000 of them annually, and their reproduction, according to available evidence, will maintain (sustain) them at the 1 million mark. However, we do not have 1 million whales – all we know is that we have a fraction of the former numbers. This does not, he said, stop mathematical modellers from saying: 'we have worked out how many you can kill per annum to keep the current numbers at *x* or *y*', but they do not know how many whales there are, and they have to work on supposition. Kleinschmidt then explained that his argument was twofold: first, that the degree of human engineering of biodiversity is wrong, because we work from the low base created by whalers over the past 150 years. 'Why', he asked, 'is this low base good or correct for our environment?'. Second, that mathematical modellers rely on 'figures' that are questionable, as, in contrast to animals on land, our knowledge of numbers (and much else) in the oceans remains scant (Kleinschmidt, pers. comm., 2008).

The CBD goes beyond CITES by establishing objectives for the comprehensive preservation of biological diversity. The CBD has three objectives: 'the conservation of biological diversity, the sustainable use of its components and the fair and equitable sharing of the benefits arising out of the utilisation of genetic resources'. The CBD also goes further than CITES by seeking to protect ecosystems and habitats, making provision for identification, monitoring and protection of areas important to species that might be endemic, threatened, migratory or have scientific, cultural, economic, social or evolutionary importance. Ideally, then, if CITES and the CBD could work successfully and in tandem, there would be international legal protection offered to endangered (and other) species both within nations and in transition between states. Certainly, there is currently much cooperation between CITES and the CBD, as between these and other major biodiversity-related conventions, but the creation of formal synergies remains in its early days (Yeater, 2013).

A fairly typical modern view of the ICRW is that it 'has become too far removed from the realities of current political needs to meet adequately the needs of whale conservation', with the reason for this being suggested that the Convention 'has survived in its current anachronistic form' simply because its members 'are divided into two deadlocked camps: pro-whaling and anti-whaling countries' and that 'any change in the balance between them may cause the machinery to collapse and so it remains static and antiquated' (Rose and Crane, 1995: 34). A question that arises, then, is whether the CBD could be used somehow to replace, to supplement or to repair the ICRW. By 'repair' is meant the bringing together of the polarised parties in common understanding and agreement as to how the Convention should operate.

Birnie writes that 'the Biodiversity Convention has introduced broader environmental and ecological perspectives into marine species conservation in general at the global level'; although claims that 'the precise problems of conservation of marine ecosystems and biodiversity have been largely overlooked by the CBD despite their inclusion within its jurisdictional scope'. On 'jurisdictional scope' it might even be asked whether the adoption of the CBD means that whaling has not arguably been brought into the fold of global governance, and Birnie asks: 'can it be said at present that a global regime (in a broad sense) exists for conservation of whales and other cetaceans, much less for conserving cetaceans as components of biodiversity or even any form of organized system?'. Birnie's assessment of this is that arguing otherwise would be to take an extremely narrow approach. Despite her concluding sentiment, the fact that states have continued to treat the IWC as the prime managing authority for whaling, and have made little or no effort to involve the CBD, implies strongly that the narrow approach is not currently debatable. Whether it ought to be debated is another question; one addressed by Birnie when she says that a 'highly contentious issue is whether there remains a compelling rationale for continuing to vest exclusive authority for whaling issues, including population assessments and assessment of threats, in a single international organization' (2003: 103, 111–2, 114–5).

Of course, it might be argued that the ambiguous nature of the CBD is simply the usual trade-off in international law – without which it might not have been possible to have had any convention agreed to at all. 'Unfortunately', writes Birnie,

> because of the many compromises required to secure consensus on an agreed text, while the CBD does provide a framework within which its parties can take the action it requires for conservation of marine biodiversity, it does not prescribe any explicit measures for doing so.
>
> (2003: 120–1)

Sand does argue, firmly, that the CBD is applicable to whaling. In his view, the CBD contains provisions which are binding on member states:

> to take measures for the conservation of biological resources, including under Article 3 to avoid environmental harm, and under Article 5 to cooperate for the conservation of biological diversity also in areas beyond the limits of national jurisdiction, as well as under Article 14 to assess the environmental impacts of projects posing significant risks for biodiversity.

He then argues that 'there is no doubt that massive killings of protected marine mammals in the Antarctic and the North-West Pacific would fall under these provisions', and that a state would not be able to avoid this by 'invoking the primacy of IWC provisions' where there is 'serious damage or threat to biological diversity'. He concludes, however, that the only recourse which CBD members have, where another member is in non-compliance, is by way of conciliation proceedings, per Article 27 (Sand, 2008: 60).

One of the most significant ways in which the CBD might come to play a vital role lies in the increased use of marine sanctuaries. Provided for in the original ICRW text, sanctuaries are not a new idea – but they have been underutilised. Probably this is because they have come to be seen by warring parties as a political tool. Properly used, sanctuaries have the potential to provide the best scientific data for us to understand – and to protect – biodiversity. It is unfortunate that sanctuaries have become so politicised within the IWC.

There is increased recognition of the need for synergies (Honkonen and Couzens, 2013). Donovan told me, however, that there is at present little cooperation between the IWC and the CBD, although 'there has been talk about it'. He then said that he 'suspected' that such cooperation 'would be more at the scientific level'. I put it to him that the CBD might be the one international instrument which could be used to bring the IWC's 'warring' parties together, but Donovan responded that he did not think that there was another treaty that could do this, given that the parties are so polarised (Donovan, pers. comm., 2007).

Until such time as the CBD and other new treaties have proved to be effective in reconciling the widely divergent approaches to conservation held by important international actors, if this optimistic prediction is ever confirmed, there will be tension between the protagonists of preservation and those who advocate sustainable use. In international law, this tension will be most visible in the inadequacies of those treaties through which countries currently seek to express their environmental views. In a sense, perhaps, what is needed is what I call an 'ecosystem approach' to international environmental law – an approach which recognises that no treaty can be truly 'stand-alone' and that, given how our system of international instruments has evolved, no one environmental issue-area can be effectively covered by one treaty only.

According to Stone, 'some parties see the framework as exploitable as a public forum for the airing of grievances on a collection of issues of deep concern and inconsistent understanding'. Another, 'overlapping' group, he argues, 'has hopes of turning the CBD into a sort of umbrella framework for all biodiversity-touching treaties and agencies', and that this latter faction sees the existing institutional landscape as being too fragmented and 'piecemeal', and therefore requiring 'coordination or consolidation under the CBD'. The governance, he adds, 'of each new multilateral environmental regime has to be built from the bottom up . . . long, frustrating, often windy start-up periods are the rule' (Stone, 1996: 129, 130). This is not just a danger. After the more than 20 years that have passed so far since the creation of the CBD, it is the reality.

Arguably, the adoption of the CBD can be seen as the most important step that we have taken – in international law – toward recognising the limits of our traditional 'categorisation of species' approach to use and protection of wildlife. This is the first global convention to urge the protection not just of particular species but the protection of habitats and ecosystems in order to provide particular species with protection. The CBD is problematic – particularly because the obligations which it places on its parties are so weak – but it is a giant step forward. Ultimately, it might well be under the banner of the CBD that the resolution to the 'preservation'/'conservation' conflict will be found.

11 ICRW/IWC membership

Discussion of membership of the ICRW/IWC means discussion of one of the most contentious issues in the context of whaling – and arguably even in international law generally. Allegations of blackmail, bribery, bullying, coercion and underhand tactics fly. In the initial decades of the IWC, adherence by new member states occurred slowly. The current moratorium on commercial whaling was the spur to the first of several rapid expansions of membership. Allied to the adherence of new members are questions of procedure and how voting should be done. These questions are not insignificant and merely procedural; the divisions and disputes over them go to the very heart of the impasse between pro- and anti-whaling states – and to the possibility of resolution of this impasse.

Many countries which have historically had no interest in whaling are members. The question must be asked why a country like Mali, to take a random example, chooses to join? To throw money away? To make enemies by stepping into a stormy debate in which it has no interests at stake?

At IWC 51, 1999, held in Grenada, Japan suggested in its (written) opening statement that 'we from Japan feel very close to the people of the Caribbean who, like us, live on small islands and depend upon the sea for much of their food resources'. Caribbean countries, Japan then argued, 'are assuming an important role in the management of the world's living marine resources' and 'their contributions deserve recognition and respect' (IWC 51, Opening Statements). Chasek and Fry have both suggested that small island states are beginning to assert themselves and to try to present a unified negotiating bloc in multilateral environmental fora. As Chasek writes, 'international environmental negotiations are likely to be a complex mix of conventional bargaining and coercive pressure applied by the more powerful actors', but that, for small island countries, coalitions – such as that represented by the Alliance of Small Island States (AOSIS) – may help to increase their bargaining power (2005: 126). Fry has described small island developing states as being 'quite an eclectic group of countries'; with there being no internationally agreed definition for them, some of them not being small (Cuba, for example) and some not being developing countries (Singapore, for example); and then makes the interesting contention that small island developing states are particularly keen on binding legal arrangements and regulation, ascribing this to their high vulnerability to factors beyond their control (2005: 89). This implies that small island states may have reasons related to deeply-held environmental concerns, rather than to pure short-term environmental interest, for IWC membership. Many observers remain cynical, however, and it may also be that, at least in the IWC context, it is a coalition that might crack quite easily under pressure brought by developed Western countries – in particular, from former colonial powers like the UK and France.

It has been suggested by Friedheim that the coalition of states which support a controlled resumption of commercial whaling is the smaller coalition, and it is generally a politically weaker grouping, with only two countries, Japan and Norway, 'being major players in world politics'. He then suggests that, despite occasional cooperation, this group has rarely been able to present a 'united coalition strategy'; Norway in particular, he suggests, has been 'criticized by some of the others for going its own way and taking care of its own interests without regard to the needs of other coalition members' (Friedheim, 2001b: 15).

Andresen argues that, while in a difficult position, the US remains one of the most important actors. As a whaling nation, he says, the US 'cannot use the same moral arguments against whaling as most other anti-whaling nations can'. He also argues that, 'as the remaining superpower, the US, at least in principle, should be concerned with the orderliness of global governance'. In conclusion, he argues that the role and interests of the US are therefore quite different from most other anti-whaling states, with interests being 'essentially monolithic in most anti-whaling countries', but in the US being broader and more diversified. Moreover, he adds, the US 'is the only actor capable of changing the present rules of the game' (Andresen, 2001: 251–2, 253–4). This assessment might be overly categoric. The argument can be made that the world is changing around the IWC as more countries move toward philosophies of sustainable usage. Komatsu and Misaki write that 'a significant development in the past 20 years, is that some of those nations that originally backed the moratorium now stand clearly in favor of sustainable whaling', and that, in addition, 'a number of nations that support the principle of sustainable fisheries have since joined the IWC and support Japan' (2003: 118–9).

There was a sudden surge in membership at the beginning of the twenty-first century. What this does tell us, if nothing else, is that whaling is not something in which few people are interested – in fact, it is an extremely active debate. This might be seen as strange, given that not many states would appear at first glance to have direct stakes in whaling. Recent joinings have included states such as Cambodia, Guatemala and the Marshall Islands, as well as new European Union members like Croatia, Cyprus, Greece and Slovenia. In 2007, even more states joined – including Israel and Laos. There were 76 member states by IWC 59 in 2007. Since then, even more have joined – such as Lithuania and Romania in 2008 – and by 2013 there were 89 members. In January 2013, however, Greece gave notice of withdrawal (probably as consequence of the depressed state of the Greek economy), to take effect from mid-2013.

For the first three decades of the ICRW's existence, its membership was small and it was definitely a 'niche' treaty. The treaty came eventually, however, to attain an importance that belies this original status, probably because of the importance of whales as an environmental symbol. At present the ICRW is in an odd position in international law – careful efforts are made to confine it, and not to allow its jurisdiction to overlap with other conservation MEAs. States are careful – vide, for instance, provisions in UNCLOS, the Convention on the Conservation of Antarctic Marine Living Resources and the Madrid Protocol to the Antarctic Treaty excluding whaling – to keep whaling as a separate 'issue area'. This cuts both ways. As the pro-whaling bloc has been careful to retain its rights when new MEAs have been agreed, so the anti-whaling bloc has also been at pains to keep CITES from exercising any authority over cetaceans.

The 'membership race' has been a dominant feature of the ICRW for the last two decades, and whether it has been positive or negative overall is a difficult thing to assess. On one hand, it may well be that the race to garner support (often from states which can have had no real interest in the matter) has contributed to much of the paralysis – the concentration on numbers has made it difficult for any states that might have liked to break the deadlock to act. When numbers are so delicately balanced, a state that tries to move into the middle risks alienating both camps. It is arguable, too, that the 'numbers race' contributed to bringing both 'bribery' and 'bullying' into the ICRW. Further, some states might have used the ICRW as a forum for political advantage, instead of contributing meaningfully to debate.

On the other hand, the argument can be made that the more that states take an interest in MEAs, and the more that all states begin to realise how interconnected all species and ecosystems are, and how these affect the lives of humans, the better. Ultimately, the bringing in of new interests, and the battle over democratic means, may help to bring greater understanding – and ultimately, perhaps, a political and legal resolution under an MEA such as the CBD.

One of the most contentious aspects of the way in which the IWC currently operates is the frantic scrabbling for support which has seen numerous states with no apparent interest in whaling adhere to the ICRW. This has then contributed to the impasse in which the states parties now find themselves. The gulf between the two camps yawns extremely wide, and this can be seen in the fights over democratic principles as well as over the utilisation of whales. This conflict has become a feature of CITES CoPs also. Although many of the new parties bring their own hidden agendas, lobby groups and political niche interests, one might hope that, eventually, the broadening of membership will contribute to resolution.

12 Special animals and links between species

It has been claimed that 'whales speak to other whales in a language that appears to include abstruse mathematical poetry' and that 'they have also developed interspecies communication with dolphins' (D'Amato and Chopra, 1991: 85). It is difficult to know what to make of the account offered by Watson in his book *Elephantoms*, where he claims to have witnessed direct communication between a blue whale and the last remaining elephant in the Knysna Forest on South Africa's coastline (2003: 170–2). As blue whales are rare anyway, and the chances of seeing one so close to shore at the same time as a single, particular elephant are even smaller, the description is probably defensible as allegory, as introducing a version of fiction's magic realism into science in order to illustrate an argument. One bit of apparent support, however, appears in a source which preceded Watson's passage – the writing of I and O Douglas-Hamilton in the early 1990s, where they say of the Knysna Forest that 'sometimes great whales could be seen undulating through the blue waters, and on rare occasions, it was said, an elephant would emerge from its hiding-place and stand in a clear patch, facing the sea' (1992: 71–2). A further snippet of support might be found in MacPherson, who writes that 'one day out of South Africa, a pod of seven short-finned pilot whales broke the surface in rolling seas. They spouted and raced beside the ship in clear sight of the elephants on deck', and 'excitedly the elephants extended their trunks over the rail and called out with guttural sounds'. As one of the whales, he says, 'rose up on a wave and came nearly within the reach of the elephants' trunks, Owalla let rip a trumpet blast . . . These two endangered species were actually *talking*' (MacPherson, 2001: 221). It is certainly an intriguing idea – communication between the two global icons of conservation. It is not unthinkably difficult to conceptualise, either – both species have shown themselves to be communicators at a level too sophisticated for easy human understanding.

In October 2012, it was reported that new studies on old recordings of the vocalisations of a beluga whale, named 'NOC', kept in captivity in San Diego for 30 years (dying in 1999), showed that he had been imitating human speech in recordings made in the early 1980s. NOC's speechlike sounds were made in frequencies 'several octaves lower' than would be normal for a beluga (Scales, 2012). By coincidence, in November 2012, it was reported that 'Koshik', an Asian elephant in a zoo in South Korea, was able to imitate human speech – deliberately putting the tip of his trunk into his mouth and moving his lower jaw in order to make sounds that an elephant could not normally make. Koshik apparently has developed a six-word vocabulary, including the Korean words 'annyong' ('hello'), 'aniya' ('no'), 'anja' ('sit down') and 'choah' ('good') (Fischer, 2012). While it was not suggested in the reports that either NOC or Koshik understood the words, and were using them to communicate appropriately, both cases represent amazing communicative abilities and researchers expressed their wonderment.

Chadwick notes of elephants that through size and the 'resulting dominion, the release from the necessity of having to look constantly over a shoulder, in a sense freed the species to develop other aspects of its existence, such as its social relationships and intelligence' (1992: 423). According to Payne, the scientist credited with the discovery that elephants communicate with infrasound, she asked herself the following questions: 'is this what I was feeling as I sat beside the elephant cage? Sound too low for me to hear, yet so powerful it caused the air to throb? Were the elephants calling to each other in infrasound?'. She then explains that 'earthquakes, volcanic eruptions, wind, thunder, and ocean storms – gigantic motions of earth, air, fire, and water – these are the main sources of infrasound, sound below the range of human hearing, which travels huge distances through rock, water and air', and that 'among animals only the great fin and blue whales were known to make powerful infrasonic calls'. No land animal, she says, was known to approach 'the power of these great mammals of the sea, but now I wondered: might elephants, too, be using infrasound in communication?'. According to Payne, the results of her experiments indicated that elephants hear and respond to each other's loud calls from distances as great as 4 km and 'an area of at least fifty square kilometres would be filled by such a call at a level that a listening elephant could hear' (Payne, 1999/8: 21, 76). Poole, writing of the process by which researchers came to understand this aspect of elephant communication, says that 'if elephants were using infrasound at high sound-pressure levels, many of our interpretations of their social world would have to change'. For example, she says, 'while we defined the members of a group visually, the elephants probably used sound, and if they could communicate with one another over several kilometers, their understanding of group composition was probably quite different from ours' (Poole, 1996: 117).

The similarities between the lifestyles of the two species would seem to offer support to shared communication characteristics. According to Payne, 'the social system that has evolved in elephants is oddly like that of sperm whales', and their 'suite of characteristics', she says, 'separately evolved on land and in the sea, has produced a pair of particularly well-adapted species' (1999/8: 117–19). Whitehead lends support to the argument that sperm whales and elephants have similar social lives, explaining of sperm whales that 'while on the breeding grounds, males rove singly among groups of females, usually spending just a few hours with each, but sometimes revisiting a particular group several times over a period of days' (2003: 449).

On increased intelligence, the reasons therefor and female influence, Payne suggests that 'a number of authors have linked social complexity to large brain size', and some have 'correlated social complexity and large brains with a slow life history, a necessary precondition for the development of both'. Old females, she says, 'become repositories of ecological as well as social information, and take on different social roles from those of younger relatives'. The learning that allows for these changes, she continues, 'occurs in the context of overlapping generations whose members spend much of their time in close proximity', with a large capacity for memory supporting 'cumulative learning'. At the same time, she says, this large memory capacity 'results in a high level of behavioral variability among individuals', and 'thus social complexity compounds itself in a network of temporal, spatial, and cognitive dimensions, each of which is, by virtue of its interactions with the others, flexible' (Payne, 2003: 85).

Further on possible reasons why elephants might have become particularly intelligent, Gavron tells us that according to Joyce Poole, the trunk is:

> such a fantastic thing. You can do anything with it. You can pick up a tree or a crumb, touch a friend with it, throw dust on your back, splash mud, toss a person thirty yards, carry a dead baby

and the trunk 'has allowed the elephant to investigate the world, to bring it closer, to look at things, play with them, use them' (Gavron, 1993: 46). Poole herself writes that the trunk 'is at once a terrifically strong and yet highly tactile and sensitive appendage and is perhaps more versatile than a human hand', and elephants use their trunks 'to eat and to drink, to mud-splash and to dust, to comfort and to reassure, to fight and to play, to smell and to communicate' (1996: 129).

Parker writes that 'I saw abundant evidence that living elephants are peculiarly attracted to dead elephants' tusks, picking up, fondling and carrying them – often for hundreds of metres'. It is difficult, he says, 'not to be subjective about why elephants do this and why they select them in preference to a skeleton's more abundant bones', and, 'seemingly they have an aesthetic sense about ivory' (Parker, 2004: 77). Meredith writes that 'elephants attach a particular significance to death' with the 'loss of a family member, especially a matriarch', being 'marked by displays and rituals which some field biologists liken to a funeral' (2001: 183). Payne describes events witnessed after the death of a young elephant, noting that on the day of death 38 different elephants made 56 visits to the body, and on the next day 54 elephants made 73 visits (Payne, 2003: 81–2). Gavron contends that 'an animal that understands death must also have some concept of life' and 'must have some awareness that it is alive and that it may die' (1993: 59). Poole agrees, arguing that 'there is something eerie and deeply moving about the reaction of a group of elephants to the death of one of their own', with it being 'their silence that is most unsettling' (1996: 153).

Speaking for myself, I have had two close experiences of elephants doing exactly this. While working as a field guide in Zambia in 1997, I was on a vehicle which was deliberately chased away, by several older elephants, from the carcass of a young, dead elephant. The carcass had been there for several days with no sign of attention from any other elephants, yet a group of elephants had come, after having paid no obvious earlier attention to it, to surround it – apparently protectively. The second (less dramatic, but no less interesting) experience came in April 2006 on a walking trail in the Wilderness area of iMfolozi Game Reserve. Here I had the opportunity to see how an elephant visited the carcass of a dead elephant, spent time there and moved the skull and several bones at the site. As a human, I have the ability to recognise a human skull; but do I have the ability to recognise it as having been a person I once knew, a skull out of place and context, as elephants may be able to do? What must it be like to be so closely connected to the world around one as to be able to do that?

The intelligence of elephants means that a number of difficult moral and ethical questions need to be considered and answered. Poole explains that she has no doubt that elephants 'have conscious thoughts and a sense of self'. If elephants, she asks,

> are conscious, thinking animals, where do we draw the line in our management policies? Is it ethically acceptable to cull entire groups of elephants? Is it ethically acceptable to leave babies tied to their dead mothers as the butchering process takes place? Should elephant sport hunting be discouraged on the basis of what we know, or encouraged because it brings in revenue that can be used to 'conserve the species'?
>
> (Poole, 1996: 158)

Sheldrick contends that elephants and humans have many things in common. They 'pine and mourn', she says, 'just as we do', and they 'can also display their deep feelings of compassion and sometimes this will even extend to other creatures in distress'. Sheldrick then goes even further, suggesting that elephants 'have also been endowed with many of the attributes we humans lack: telepathic powers of communication, the ability to reach across the miles through infrasound below human hearing range' – and even 'that mystical genetic memory' we call 'instinct'. Like humans, she adds, 'they also have to learn, and the wisdom of the elders is passed from one generation to the next' (Sheldrick, 1997: 26–7).

Meredith writes that 'the one factor that has moved decisively in the elephant's favour in recent years is the work of field biologists', with their discoveries having 'revealed an ever more remarkable and complex animal' (2001: 225–6). Poole contends that 'teaching, which has been assumed to be a uniquely human attribute, is one behavior we have observed in elephants'. She then explains that 'when people ask what I find special about elephants, one of my answers has always been their sense of humor'. To have a sense of humour, she argues, 'one has to have not only an ability to see the funny side of a situation, to see a situation in a larger context, but also a sense of self' (Poole, 1996: 139). The question then must become, of course, why might this not be true for more species than just elephants and whales? Why not those remarkably intelligent animals, pigs?

Of whales, Martin writes that some researchers, 'such as Jacques Cousteau, see the changing complexity of humpback whale song, and the fact that the songs are recognised and changed by other males of the stock, as evidence of a form of language'; while others 'are not as sure and are not willing to see it as anything more than a stylised display' (2001: 183–4). Payne, *et al.* have suggested that 'detailed analysis of recordings of humpbacks provides evidence that the changes in song are not simply due to forgetfulness between singing seasons'. They explain that most changes, in their research, did not occur seasonally, but instead occurred 'during the time when the whales were singing, developing their songs methodically in measurable steps'. Furthermore, they say, 'the types of change varied from season to season, and so could not be attributed to repeating seasonal factors'. They make the point that they 'know of no other animal where whole populations introduce such complex, rapid and non-reversing changes into their vocal displays, abandoning old forms and replacing them with new'. The same writers then argue that the progressive changes in whale song can 'thus be seen as a form of cultural evolution, in the sense that the song is a learned trait which evolves' (Payne *et al.*, 1983: 11).

These contentions are supported by Guinee, *et al.*, who argue that their research has shown that 'individual humpback whales change their songs over time and that the changes occurring in the songs of individuals tend to parallel the changes made in the population as a whole'. They argue that the 'probability that whales modify their songs in the same way at the same time by chance is staggeringly small, and we must conclude once again that they listen to each other and learn the changes in the song'. Their comment on this is that it is 'most certainly, another clue into the mental capabilities of the species' as, 'in order to learn and adopt the current version of the song at all times, humpbacks must continuously be attending to a very large collection of details' (Guinee *et al.*, 1983: 78). Far more difficult than ascertaining that whales communicate, of course, is trying to work out what it is that they are saying *to* each other. Clark writes that in study of southern right whales, the complexity of the social context was shown to be directly related to the complexity of the sounds made, with this suggesting 'not only that the whales are communicating acoustically but that they are using sounds to communicate a complex array of messages' (1983: 163).

By way of partial explanation of whale intelligence, Whitehead argues that 'most animals are both predators *and* prey', and that 'even the very large are vulnerable' with, for the sperm whale, killer whales being 'a real threat'. The structure of the oceanic habitat, argues Whitehead, 'has consequences for confronting such threats' as there 'is nowhere to hide, and the three-dimensional habitat structure makes defence against predators, and perhaps also vigilance, more difficult than on land'. In this regard, it seems that 'social structures can help, and communal vigilance for predators and communal defense against them have both been considered as important functions of the social structures of pelagic animals' (Whitehead, 2003: 448).

As an interesting link between species, both whales and elephants have been recorded exhibiting behaviour that seems to show sympathy and concern for other species – even to the extent of 'rescuing' individuals from other species. As recently as March 2008, for instance, an incident was reported from New Zealand of a bottlenose dolphin leading two distressed whales to safety in deep water. Rescuers were attempting to persuade two pygmy sperm whales, which had repeatedly beached themselves, to leave – they were tired and apparently in deep trouble when a resident dolphin seemingly deliberately led them to safety (Project Jonah, 2008). Conservationist Lawrence Anthony, who owned a game reserve in KwaZulu-Natal, wrote of an incident where an elephant matriarch was witnessed opening a gate in order, apparently deliberately, to release from captivity in an enclosure a group of about 30 nyala antelope (Anthony, 2009).

However, there are commentators who dismiss the 'special nature' of elephants and whales. Friedheim argues that a distinctive feature of the problem of managing whales – and elephants – is that many people in developed countries think of them as being 'special creatures' because of their size or purported intelligence 'and therefore assert that they require a special regime not necessarily congruent with regimes for other forms of wildlife'. But, he asks, ought these species be exempted from the normal rules developed to achieve sustainable use? (Friedheim, 2001b: 25).

The differences of opinion are vast. M'Gonigle, for instance, writes that 'many now believe that cetacean intelligence, although of a profoundly different character than that of a man, is man's rival' (1980: 121). Lapointe, on the other hand, records that the sperm whale has the largest of all mammalian brains, but size, he says, 'does not equal intellect' and whales 'are approximately the intellectual equal of ferrets (toothed whales) or chickens (baleen whales)'. In support of this allegation of poor levels of cetacean intelligence, he offers that 'their cerebral cortex is relatively less developed', and that whales 'also do not have REM or Rapid Eye Movement, which is a characteristic of most land animals' (Lapointe, 2003: 81).

Broch contends that animal rights' defenders 'anthropomorphize the whale' so that 'it becomes at least as intelligent as humans and it is nurturant and has developed a sense of humour, and is believed to want to befriend and protect humans' so that, in a way, 'the whale represents innocent, good, unspoiled human qualities' (1994: 212). Kalland writes that 'in Lévi-Strauss' terms, whale society has become a metaphor for the lost human paradise or utopian world and caring for whales has become a metaphor for kindness', for being 'good', and that 'the anti-whalers create a totemic dichotomy of mankind, with whales as the totem for themselves and with money as the totem for the whalers, a world view strongly opposed by the whalers' (1994: 163–5). Ris agrees, arguing that 'by communicating the whale as a totem, it is possible to distinguish visible and comprehensible opponents, i.e., whalers' (1994: 224).

Ultimately, it might be that proper protection of biodiversity can be argued to require that more animals than elephants and whales be accorded respect and even rights, but with elephants and whales being symbols of this. Darby quotes Thoreau as asking: 'can he who has discovered only some of the values of whalebone be said to have discovered the true use of the whale? Can he who slays the elephant for his ivory be said to have "seen the elephant"?' (2007: 100). In my opinion, the answer must be a resounding 'no'. As Martin writes,

> as whales continue their passage across our collective mind, they reaffirm a growing view of humans living in a world of shifting and fragile ecological

> balances, where resources are shared with a multiplicity of species. All are dependent on a secure ecological place and the effects of human change.
> (2001: 233)

Perhaps a neat way to conclude this discussion on the common features (and shared destinies) of whales and elephants is by pointing to parallels drawn in one of the most famous ever works of literature – *Moby Dick.* Herman Melville draws comparisons several times. He refers to a particular species of whale as an 'Elephant Whale'; in what many consider to be the seminal chapter of the novel (Chapter 42: 'The Whiteness of the Whale'), he draws comparisons with ancient and modern rulers elevating above all other standards that of the white elephant; he refers to a whale rising beneath a whale boat as making 'an enormous wallowing sound as of fifty elephants stirring in their litter'; and he compares whales seen from the masthead as being reminiscent of 'recumbent elephants' on the plains of India (Melville, 1851/1994: 149, 189, 224, 269).

Melville next suggests that 'in the whale the sense of touch is concentrated in the tail' and that 'in this respect there is a delicacy in it equalled only by the daintiness of the elephant's trunk'. 'Had this tail', he adds, 'any prehensile power, I should straightway bethink me of Darmonodes' elephant that so frequented the flower-market, and with low salutations presented nosegays to damsels, and then caressed their zones', and:

> on more accounts than one, a pity it is that the whale does not possess this prehensile virtue in his tail; for I have heard of yet another elephant, that when wounded in the fight, curved round his trunk and extracted the dart.

He then describes the comparison of tail and trunk as being one of 'chance' and says that this 'should not tend to place those two opposite organs on an equality, much less the two creatures to which they respectively belong'; for, he says, 'as the mightiest elephant is but a terrier to Leviathan, so, compared with Leviathan's tail, his trunk is but the stalk of a lily' (Melville, 1851/1994: 361, 362, 363).

Finally, Melville suggests – perhaps somewhat unfortunately for this book – that 'all comparison in the way of general bulk between the whale and the elephant is preposterous, inasmuch as in that particular the elephant stands in much the same respect to the whale that a dog does to the elephant'; but Melville rescues my efforts by adding that 'there are not wanting some points of curious similitude', giving as an example 'the spout'. 'It is well known', he says, 'that the elephant will often draw up water or dust in his trunk, and then elevating it, jet it forth in a stream' (Melville, 1851/1994: 363).

The aim of this book has been to demonstrate that links and similarities between elephants and whales – and the real importance of these links – go far beyond the ability to spout water, and that these links have the potential to influence, or perhaps even to determine, the future course of humankind's relationship with the two species – and with all other species. It is no accident that elephants and whales have become the two most iconic species of the world's environmental movements, and, indeed, a very battleground upon which the future of the way we live with animals might be decided. Both are remarkably complicated, sensitive and intelligent species. We have made incredible advances in our understanding of the complexities of animal behaviour, and of the ethics of our abuse, consumption, interactions with, management, stewardship, trusteeship and use of animals; but we still have a long way to go.

13 Toward a conclusion

At the end of more than 15 years of thinking about, and ten of writing on, this subject, concluding is no easy task. In one sense, this book is very simple. It is a gathering of evidence (direct and indirect) to support a contention that there are links that have gone largely unrecognised between whales and elephants, the ICRW and CITES, and to collect such evidence in writing and thereby to make a modest contribution to knowledge. It might be objected that the idea is simple, even trite – but I did not find anyone else who had done it and so I hope that I have made a contribution. Of course, more is needed than mere collection if the reasons for, and the significance of, these links are to be understood.

13.1 The battle and the war

This book has examined linkages between two multilateral environmental treaties, and linkages between two species. In order to understand the linkages, it has also been necessary to understand the histories of governance of the two species, their histories within both national and international law and something of the various human participants (be these nationalities, groups or individuals) who make decisions in respect of the two species, or who are affected by such decisions. The world's understanding of management of natural resources is changing, and the politics surrounding issues may overlap and influence other issue areas.

This book has dealt with the nature of multilateral environmental treaties, through discussion of the ICRW and CITES, their histories, the ways in which they have been used and abused by their parties and the links between them. Multilateral environmental treaties are affected by changing mores and *zeitgeists* – becoming ever more difficult to use effectively as the world around them changes. In the case of the ICRW, the treaty was concluded as the culmination of pre-war efforts to settle control and administration of the use of a resource – and the treaty is arguably too rigid to be appropriate for the twenty-first century. Unfortunately, however, its parties have become, over the years, so locked into opposed positions that it is on the one hand almost impossible to conceive of a new treaty being drafted, and, on the other, equally difficult to imagine the gap between them being bridged – as such, we limp on with an inadequate management instrument. CITES is a newer treaty, adopted in an age of different environmental sensibilities and as part of a cluster of similar treaties, and is more flexible and reflective of wider interests – but it is also beginning to show signs of middle-age (tiredness, expansion of girth, weakening eyesight and lack of focus).

Of course, both treaties could arguably be made to work, and to work well – if their parties were ad idem. This requires discussion of the opposed philosophies of preservation and conservation, as these terms have been used in this book. It may be that eventually the world's states will become reconciled in a common view of how natural resources should be protected, nurtured and used; at present, however, this is not possible – there are too many competing needs and interests. These competing wants manifest themselves in international law and politics in a tangled morass of compulsion, compromise, bullying, horse-trading and secret deal-making. The sovereign status of statehood hangs always above – like an albatross-shaped millstone of Damocles for those seeking to bring states together to protect the environment, for present and future generations and also for its own sake.

As Hegel puts it, the:

> fundamental proposition of international law (meaning the universal law which ought to be absolutely valid between states, as distinguished from the particular content of positive treaties) is that treaties, as the ground of obligations between states, ought to be kept.

'But', he continues, 'since the sovereignty of a state is the principle of its relations to others, states are to that extent in a state of nature in relation to each other'. 'It follows', he then adds, 'that if states disagree and their particular wills cannot be harmonised, the matter can only be settled by war' (Hegel, 1821: 213–4).

The world has still not awoken to the damage done to the environment by the insistence on freedom to act as one pleases on the high seas – little did the Arbitral Tribunal deliberating on the fate of the Bering Sea fur seals know what a destructive precedent it was setting when it ruled. Paradoxically, the world has not realised, either, the damage done to the environment by according states complete autonomy over natural resources within their artificially constructed borders. Little, either, did the parties to the Paris Convention understand that one day the bearded vulture they had classified as a 'noxious' species would be considered a bird so rare that seeing one would become a special experience. There are some inklings in international law of new understandings in this regard – most visible perhaps in the CBD – but it will be many years yet, and perhaps not even within my lifetime, before it is understood that neither of these extreme positions, used to justify the pillage of the oceans and the sack of environments everywhere, can be justified. The saddest thing of all, in my view, is that the damage which has thus far been done to the world's animals (wild and domesticated), and to its environment generally, is merely a scratching of the surface compared to what still needs to happen before people will recognise the real sacrifices they need to make – both nationally and internationally.

In early 2013, the ICRW had 89 (soon to become 88) states parties and CITES had 178. They are significant treaties. The arguments at their meetings are incredibly bitter, and this is because so much is at stake. What I have tried to suggest in this book is that these two treaties – and the two iconic species over which their parties fight – have provided the battleground for one of the most important battles of the human war for survival. It took the deaths of unknown millions of these two species even for them to become that battlefield – and when one thinks too of the untold (uncounted and uncountable) billions of other animal, bird and fish species which have suffered and died for or because of humanity, it is tempting to recast well known words and write that never have so few owed so much to so many.

Many states have been involved in the battles described here – and few have emerged with their standards shining. The long history of human usage of the products of the whale and the elephant has been a history of shame, hypocrisy and self-interest. Few states have been consistent, and many have switched allegiances – sometimes temporarily and sometimes apparently permanently.

The importance of this battle within the greater war cannot be underestimated – what happens on this battlefield will sway what happens in the end, whether we come to realise it now or only much later. According to Hegel,

> it is as particular entities that states enter into relations with one another. Hence their relations are on the largest scale a maelstrom of external contingency and the inner particularity of passions, private interests and selfish ends, abilities and virtues, vices, force, and wrong.
>
> (1821: 215)

It seems to me that this would be a fairly accurate description if applied to the international regimes to which whales and elephants are subject.

13.2 In my view

I began this conclusion by suggesting that this book has examined linkages between two treaties, and between two species. In researching these links, I found many unexpected connexions. Sitting with records in the archives at the IWC Secretariat and discovering that, for four consecutive years in the late 1990s, Zimbabwe had attended IWC meetings and had presented (written) opening statements each time – and in each one had discussed elephants, rather than whales – was a real find. Being told that the Japanese Ambassador to South Africa had asked for support on whaling, in return for Japan's support on re-opening the ivory trade, was another such moment.

The research experience has been incredibly valuable for me. Many times over the years I have been frustrated and felt that my chosen subject was so vast that I would never be able to get a proper grip on it. However, the exhilarating feeling that I was beginning to understand the debate as I read more and more, and as I spoke to more people, and as I travelled to countries I might otherwise never have visited, has been wonderful. In particular, having the opportunity to join the South African delegations to IWC 59, 63 and 64 in Anchorage, St Helier and Panama City gave me insights into the process that I would not otherwise have had. Beyond that, the friendliness showed me by people involved in the debate – from Cape Town to Tórshavn, Nairobi to Anchorage – and their willingness to discuss the issues with me, gave me hope that it will eventually be possible for the world to accommodate different views.

I visited the Faroes in 2007, in order to research the '*grind*' – the driving of pilot whales (and occasionally other species) onto shore for slaughter. Faroese pilot whaling does not fall under IWC management control and the Faroes takes pilot whales legally. I did not, unfortunately, get to see a *grind* – it being random as to when the whales might come near enough to be driven ashore – but I did get to interview a number of people involved. I was hosted by the Head of the Faroese Pilot Whalers' Association, and also interviewed various researchers and the veterinary surgeon most closely involved with improving killing methods.

To the Faroese this is an essential part of their history and culture – so that even though they are a modern society with no need for the meat as food,[1] they continue the practice. It is so important to them that when a pod of whales is sighted and the alarm is given, people involved drop whatever they are doing at work and rush to the relevant spot. (There are 18 islands in total, but they are close together.) The meat is then allocated, entirely free of charge, to any people/households who want some. It is never sold. There are less than 50,000 inhabitants.

Probably about 800 to 1,500 pilot whales are taken annually from a total population that might be as high as 800,000. While unable to comment on the killing methods in practice, I was informed that it all happens remarkably quickly[2] – the figures which I was given certainly compared very favourably to the general slaughter of animals. On a field trip to a reputable abattoir in KwaZulu-Natal, South Africa, I was informed that it is legally required that not more than six slashes be used to cut the throat of a cow – after it has been stunned with a bolt gun and hoisted into the air by a chain. However, I did see several instances where the slaughter took up to 12 slashes – and even then several cows seemed not to be dead when they were taken off further to be disemboweled. While the Faroese do arguably have a 'blind spot' in respect of cetaceans, and while they do import certain meat products, I felt that in contrast it was probably the society I have seen which treats its animals better than any other – certainly far better than my own country, South Africa. In the Faroes, farm animals generally wander unimpeded, and I was informed that there was only one building in the Faroes where cattle are kept indoors. When animals are slaughtered, it is usually done quickly and unexpectedly by someone whom the animal knows well.

1 Indeed, some of the whale meat taken is so imbued with heavy metals (mercury and cadmium) and polychlorinated biphenyls (PCBs) that young and pregnant women are advised not to eat it. The Faroese generally consider it safe to eat *grind* twice a month, however. The levels of chemical contamination apparently vary substantially – probably according to where the whales have been feeding (Sjúrðarberg, Bloch, pers. comms, 2007).

2 (Sjúrðarberg, Olsen, pers. comms, 2007.) Of course, if I imagine myself standing waist deep in rough seawater alongside a thrashing 3 tonne beast and imagine trying to kill it with a precise thrust of a metal blade, I have inevitable misgivings.

While researching, I have met many truly wonderful people. The real value of conducting interviews, be these formal or informal, lay sometimes in the background and context they gave me. For instance, I interviewed Nan Rice in Cape Town – Nan has for many years studied dolphins and whales and runs an NGO, the Dolphin Action and Protection Group. In general discussion, Nan told me that in the late 1970s, in the early days of her work with cetaceans, she had lived next door to a man named Andrew Behr whose two daughters (Susan and Theresa) had played with Nan's own children. She did not find out until much later that Behr was the owner of one of the world's most notorious pirate whaling operations in the late 1970s – with two notorious whaling vessels, the *Susan* and the *Theresa*. Anecdotal, but fascinating and all part of the research matrix.

At the same time as being so welcomed by people on all sides of the debate, however, the process has been terribly depressing. Some of the opposition to opposing views runs so deep that it is visceral. It will be a very long time before there is reconciliation of these views, and many animals and people will suffer unimaginably during that time.

Many times over the years I have been asked what I think of the subject, what my own views are. It has always been difficult to explain that I am in the process of trying to understand all views and that I have been deliberately avoiding reaching firm conclusions. In the course of research, this (usually vegetarian) researcher has eaten elephant (biltong) in the Kruger National Park, minke whale (delicious in wine sauce) in Northern Norway and harbour porpoise (roast, and too strong a taste for me) in the Faroes; drunk beer at midnight while bobbling in a rubber duck on the open sea off one of the world's largest storm petrel colonies at roost after setting nets for seals (in order to fit them with radio collars); been charged by elephants (to the point that this book might never have been completed); fed embarrassingly too many fish while watching whales; changed words in resolutions of the IWC and accidentally set off an acrimonious hour-long debate on the floor in plenary session; and tried through it all to remain as objective as possible. For the last decade, my immersion in this research has sometimes felt almost total.

For what it is worth, though, I believe firmly that in the years ahead we are going to come to understand that animals are very much more intelligent and animal societies far, infinitely far, more complicated than we currently realise. Eventually, too, we are going to be forced to acknowledge that the damage we have done to the environment is immoral beyond belief. Whether we will learn this in time to save anything meaningful of it in its natural form is, I believe, highly unlikely. In the meantime, while we limp toward this understanding, the best that we can strive for is to mitigate the damage we do as individuals, and we can try to persuade our representatives in the international arena to take decisions from positions that are as informed and as moral and as humane as possible.

In international management of species, until we are able to change our ways and consume less – or no – animals, we ought to manage them as best we can in as full an understanding of biodiversity, and with as precautionary an approach, as possible. While I abhor the thought of the taking of whales for human consumption, I am sickened even more by the careless disregard with which we consume less high profile species – such as tuna. If the taking of whales were to mean less taking of other species, so that the ecosystems in which the whales lived were to remain in a better balance than with the continued unsustainable taking, I could better accept this – until such time as we recognise that neither is truly acceptable. The cumulative damage which humankind is doing to the environment generally is probably far worse than the damage being done to any individual animal.

In international law, we can work toward increased understanding of the interconnectedness of treaties and of how the decisions we take may affect other treaties, people and species in ways that might not be immediately obvious. I have never fooled myself into believing that my writing this book will solve any of the problems it deals with – the differences between viewpoints are so vast that this cannot be done except by many people in many places and with great patience. It is my hope, instead, that this book will contribute toward the understanding of the places of two wonderful species in international law and politics, of the interconnectedness of treaties and of the interdependent relationships between people, between animal species and between people and animals.

In this regard, my recommendation for the future is that we recognise that in the same way that we have moved from offering protection to species in isolation to affording protection to ecosystems, so we need to understand international conventions within their 'ecosystems' – as we are coming to understand the dependence of species on each other, so we must come to understand the connexions between treaties. Even if the world moves toward increased regional management of particular ecosystems, framework conventions like the CBD must be used to bring together other disparate instruments like the ICRW and CITES. Biodiversity depends on linkages and we need to understand the inherent biodiversity of species and the inherent biodiversity of laws – no matter how large and grey these diversities might seem to be.

14 In black and white and shades of grey

Recommendations

14.1 The problem

During the writing of this book, and the thesis which preceded it, I have always been careful not to delude myself into believing that my research would elucidate with finality, and provide neat solutions for, problems which international scholars and negotiators far more experienced than myself argue over and have not been able to solve. My object was, instead, to contribute to the debate by researching an area which had not been previously traversed in detail. In that sense, I hope that I have completed the task which I set for myself. I will, however, conclude by making certain recommendations for the future.

It has been one of the main sub-texts in this book that a core problem with both the ICRW and CITES is that they take a single-issue, species-specific approach. The reasons for this are to be found in the history of the development of international environmental law, and in the slow pace at which our understanding of how best to manage and protect biological diversity has developed. How best to move away from such an approach, and to institutionalise the linkages which have developed between the ICRW and CITES (and other treaties), is the problem that needs to be solved. At stake is nothing less than adoption of the best possible philosophy to guide the future use and protection of natural resources.

If it is considered that both the ICRW and CITES have mixed records and that both are proving problematic in implementation, and that resolutions to these problems should be found, there are essentially three directions in which such resolutions might be sought. First, a new convention, or conventions, might be drafted to replace or supersede the existing conventions. Second, changes might be made to the existing conventions. Third, a way might be found to bring the existing conventions in line with a new guiding philosophy. The first seems unworkable, given the levels of mistrust and the degrees of vested interest which presently persist. The same objection attaches to the second, although it might be easier to obtain changes than to replace wholesale. The third is probably the most viable in the present circumstances. This is the approach which it is suggested be adopted – that instead of trying immediately to make radical and sweeping changes, warring parties begin to move closer together, to find common aims, and thereby to prepare the ground for eventual rapprochement.

14.2 Recommendations

As the international environmental lawyer watches international legal provisions move, with 'catlike tread', from 'soft' toward 'hard' law, so it is imperative that precision is required in language. This will enable terms themselves to 'concretise' and for the principles which they embody to become better understood. Allied to this, careful note should be taken of the way in which discourses shift over time. In this regard, for instance, attention should be given to such terms as 'conservation', 'ecosystem', 'preservation', and 'science-based', as how these are used, or abused, can have important implications for the ways in which 'sustainable development' is to be understood.

Probably there is not anything that can be done to create trust between states quickly – at present, the enmity runs too deep. There is considerable division in opinion even on the very nature of treaties. In respect of the ICRW, for instance, some say that it is a flexible treaty while others maintain that it is overly rigid; some describe it as outdated, others as prescient of modern understanding; some see it as an environmental treaty, others disagree. There are, likewise, very different views even of what constitutes democracy in international practice.

The sooner that a process of facilitating compromise begins, the better. In recent times, it does seem that new efforts *are* being made to facilitate trust, slowly. This is happening in both the IWC and CITES. The real value of compromise is that if it can be reached in the contexts of whales and elephants, two species which have been so bitterly fought over, it can surely be reached in the specific contexts of other species. In the building of a facilitation process, an important recognition to make is that 'Pareto optimality' is an obstacle to be overcome. It should be understood that while all decision-making parties may be benefiting, it is the environment itself, and certain groups of marginalised people (including future generations, the most marginalised actor of all with the exception of the environment itself), who might be suffering or will suffer in the future. What I mean is that while the major actors on the international stage are able to press their issues, and perhaps to benefit from the lack of change, those actors which do not have voices, or which have voices that go unheard, might not have their interests considered.

It is relatively easy to make suggestions as to specific provisions that might be adopted – in the fairly sure knowledge that it is unlikely that any of them *will* be so adopted, which means that the value of the suggestions is unlikely to be tested. What I am more interested in doing is suggesting a shift in general overall approach. Nevertheless, certain specific suggestions can be made which would improve the prospects for success of such a shift occurring, and for its being successful, should it occur.

The use of reservations to provisions of international environmental treaties should be strictly limited; while it would be difficult, and perhaps even undesirable (given that their use might induce states to join a treaty which they might otherwise not join), to prevent their use altogether. It should be made more difficult for states parties to back away from, or to sidestep, commitments they have made. It is imperative that there be improvement of enforcement provisions, and that such improvement be closely coupled to improvement in reporting obligations and monitoring techniques. Sanctions should also be made more sophisticated – more 'targeted'. In respect of both monitoring and enforcement, it would be very helpful were jurisdiction to engage with wildlife trade issues *within* member states to be added to CITES's remit – at least on a limited basis. This might best be achieved, at least initially, by significantly improving formal linkage with the CBD.[1]

It must be acknowledged that there are urgent problems to be dealt with in the areas of growing under-capacity and the increasing sophistication of corruption. If these are denied, little progress will be made. Another possible step in respect of enhancing enforcement, although one which would need to be very carefully managed, would be to bring the WTO closer into the cluster of environmental treaties – in order to utilise its 'teeth' (its strong punitive provisions) to improve compliance. At the same time as improving enforcement provisions, it should be recognised that a more effective system of true incentives for compliance should be introduced. The 'carrot' and 'stick' approach will not work for long unless both are real possibilities. Of course, a hard pill to swallow is the acknowledgement that far more resources (both financial and logistic) need to be put into all of these things. Eventually, this will happen – but a 'push' now would be invaluable.

There should be recognition that all sides to the debate over sustainable use have, or at least may have, reasonable arguments to make. Almost certainly, there is value to be found within all such arguments. More importantly, however, it should be recognised that it is both vulnerable people and the environment which suffer from the imposition of misguided policies and from the regulatory vacuums created when states reach impasse.

It would be helpful for there to be recognition that law should evolve as understanding does. Even if a treaty *could* be made to work, were its states parties willing, the perception that it is flawed is itself damaging. Although, probably, we are not in a position to negotiate new treaties and must go on with those we have, we could arguably deal with this best by clustering them together under an umbrella – in which regard the CBD may currently provide the best sheltering cover we have. While the proliferation of regional treaties might appear to make this difficult, even this might be turned into an advantage so that habitats (both on land and at sea) can be managed with the best expertise available, while guiding principles gain

1 Although the CBD remains 'obeisant' (especially per Article 3: *Principle*) to the concept of state sovereignty, it takes a significant step toward breaking through the barrier imposed by this concept. Article 4: *Jurisdictional scope*, for instance, states that:

> Subject to the rights of other States, and except as otherwise expressly provided in this Convention, the provisions of this Convention apply, in relation to each Contracting Party:
>
> (a) In the case of components of biological diversity, in areas within the limits of its national jurisdiction; and
>
> (b) In the case of processes and activities, regardless of where their effects occur, carried out under its jurisdiction or control, within the area of its national jurisdiction or beyond the limits of national jurisdiction.

clarity at a much broader level. It would be useful to recognise the value of regional expertise – and to encourage regional treaties.

It is essential to recognise, as a matter of the utmost urgency, that 'freedom of the high seas' is a deeply, ***and irreparably***, flawed concept – and that so too is absolute national sovereignty, especially where species which have global impacts are concerned. We desperately need a new understanding of, and attitude toward, seas and oceans. This I see as being the greatest looming tragedy of all. We must not divorce treaties from treaties, and thereby divorce species from species – especially whales from fish. On this one, we simply cannot wait. Allied to this, the problems created by the treatment of national sovereignty as sacrosanct must be recognised, such as that this approach denies that some issues, and at least some species, must supersede national boundaries because of their global significance.

Greater use of the precautionary principle is extremely important because of the overconfidence of science – or, rather, because of the overconfidence with which political leaders appear to understand, and to make use of, science. It would be very useful to begin to include the term in important international documents – and not to mask it with the term 'precautionary approach'. The point which I am making here is that as long as state actors are reluctant to recognise an approach as having become a principle, they show necessarily a reluctance to commit to the approach as binding on themselves. Until states do begin to make such commitments, there will always be a considerable degree of uncertainty over the degree to which environmental protection should be taken into account, and it will be easier for states to avoid making meaningful commitments to such protection. The status of the precautionary principle is arguably a strong bellwether to indicate the direction in which states are heading and the fervour with which they are taking that course.

There needs to be explicit recognition of the problems caused by categorisation of species – especially where this has been formalised in international treaties. While this recognition might well be happening slowly already, through formal and informal cooperation between different conventions, more would be helpful.

The roles of local communities in protecting environments must be recognised. The interests of such communities must be brought into the ecosystems of treaties which it is proposed be recognised as a form of 'biodiversity of laws'. In this respect, as examples worthy of study, the transboundary conservation area should be recognised as an intermediate stage between local/national and international governance; conservation programmes in Southern and Eastern Africa are particularly worth considering, if not always of emulating; and Aboriginal Subsistence Whaling is an issue of importance beyond merely its local impacts on affected people and whale species. Allied to this, perhaps, the incredible value of sanctuaries and strictly protected areas should be recognised.

Following on from this, the paradox must be recognised that in certain circumstances it might be an approach more protective of biodiversity to give private individuals greater rights over areas or species, to encourage the protection (and/or sustainable use) of these; while in other circumstances holding areas or species in public trust would serve both their interests and the interests of people (including future generations) better. This requires, of course, a delicate balancing act – but an essential one. It may also be that on occasion trade in individual animals might benefit a species, although I believe that overall trade needs to decrease not increase.

It must be recognised that trade has global effects and that trade in one species in one place might have unforeseen (even unforeseeable) effects on others. A new approach could recognise flexibility – so that, for instance, a species abundant in one area might be traded while in another area the same species might be given strict protection. 'Split-listing' as in CITES may be a way to achieve this – if used judiciously. In this way, unique circumstances of trade affecting biological diversity can be taken into account. It is also important to include, in both the study of possible future courses of action and negotiation toward such possibilities, recognition of environmental challenges other than trade – the effects of climate change, habitat destruction, invasive species introduction and various forms of pollution, for instance.

There should be recognition of the inherent 'biodiversity' of treaties, and of the fact that treating them on their own is as artificial as is treating species on their own, and of the practical importance which this recognition can have. This might be an insightful way to approach compromise, especially as it leads directly toward the CBD, which, it is suggested, might provide the most appropriate battleground on which to negotiate an armistice. Ultimately, it seems to me that the CBD is the *one* current convention, intended by its very nature to provide umbrella-style protection and in itself the product of compromise, which might provide such an opportunity for 'rapprochement' of views.

In respect of seeing the CBD as an 'umbrella' beneath which to reconcile different views, it is worth pointing out that the CBD, as can be seen from a reading of its preamble, recognises all of the 'intrinsic value of biological diversity'; the importance of biological diversity for 'maintaining life sustaining systems'; a formulation of the precautionary principle; both *in-situ* and *ex-situ* conservation; and the dependence of indigenous and local communities on biological resources, and the value of the knowledge and culture of such groups. Further recognised are the importance of 'international, regional and global cooperation among' states, intergovernmental organisations and NGOs; the special needs of developing countries, least developed countries and small island states; and the desirability of 'enhancing and complementing existing international arrangements for the conservation of biological diversity and sustainable use of its components'.

There are some possibilities that frustrated pro-whaling members might eventually decide to leave the ICRW/IWC. At present it seems unlikely that this will happen; but, should the IWC collapse, it seems to me more likely that it would be replaced by a number of regional instruments than another single-issue, umbrella-type MEA. When Russia suggested, at IWC 55 in 2003, that 'whale conservation could be given high priority under the framework of the Convention on Biological Diversity', the suggestion was not supported by significant numbers of states (IWC 55, Chair's Report). I think that this was a missed opportunity.

Just as it is illogical to see whales as somehow separable ecologically from the fish species sharing their habitats, so is it illogical to see elephants as separate from the smaller species in *their* habitats. And, as illogical as are these, so is it illogical to see CITES and the CBD as separate – this must be recognised and the two must operate in harmony. In order to introduce greater linkage with the CBD, a formal step which might be taken is the introduction into the IWC and CITES of 'Joint Implementation Committees' – bodies charged with integrating the aims, the work and the processes of the three. Such Committees could be either bilateral, perhaps initially, or trilateral.

The nettle of the whaling issue needs to be grasped and the regulation and management of whales could be brought under the auspices, the guiding principles at least, of the CBD. A relatively low-key start in this regard might be a joint, decisive acknowledgement by IWC members that the principles of the CBD are applicable to the regulation and management of whales – in 2010 such a decision was considered, but not adopted (IWC 62). The objectives of the CBD are the conservation of biological diversity, which is clearly an objective also of both the ICRW and CITES; the sustainable use of biodiversity's components, which is an object that few would argue against – the fight being rather to determine what is meant by 'sustainable' and by 'use'; and the 'fair and equitable sharing of the benefits arising out of the utilisation of genetic resources', with a recognition of the importance of equity being an important step toward eventual resolution of conflict. If equity could be extended to include recognition of the rights and interests of future generations – and perhaps even eventual recognition of the rights and interests of the environment itself – significant progress would have been made.

At CITES CoP 16, in early 2013, a resolution on the 'Cooperation of CITES with other Biodiversity-related Conventions' was adopted. The CBD and the Convention on Migratory Species featured heavily, the ICRW less so but that it was included is a hopeful sign in itself. All that was said, per the Secretariat's accompanying information document, was that 'in accordance with Resolution Conf. 11.4 (Rev. CoP12) on Conservation of cetaceans, trade in cetacean specimens and the relationship with the International Whaling Commission', the:

> Secretariats of CITES and IWC have advised each other about meetings of mutual interest, revised the schedule of species and stocks protected from commercial whaling by the International Convention for the Regulation of Whaling, and otherwise exchanged information of mutual interest.
>
> (CITES, 2013)

Not much, but something to work off.

The symbolic importance of whales and elephants needs to be recognised. This is easier said than done, in that the 'conservationist/preservationist' camp is presently likely to argue that the species have inherent value, and the 'conservationist/sustainable use' camp to argue that such value has been inflated by the species having morphed into unrealistic symbols. To recognise, however, that the species may, as symbols, be representative of all other species, would be a significant step forward. Rightly or wrongly, the two species *have* become essential global symbols for different environmental philosophies, and we must make use of this – 'turn it to good'. The real point being made here is that we should recognise the statuses of both whales and elephants *as* symbols, for better or for worse, *but* at the same time we should make this recognition more sophisticated by aligning the species directly with other species and their habitats. At the same time as doing this, we must be careful not to understate the importance of both species as essential physical keystones within their various ecosystems.

There is a multilateral environmental treaty which provides, I will argue, an interesting and important precedent for how the guiding philosophy, and consequently the operation, of a treaty might be transformed. This is the 1972 Convention on the Prevention of Marine Pollution by Dumping of Wastes and Other Matter, also known as the London Dumping Convention (or just the 'London Convention' in context). In this Convention, 'dumping' is the 'deliberate disposal at sea of wastes or other matter from vessels, aircraft, platforms or other man-made structures, as well as the deliberate disposal of these vessels or platforms themselves'. The approach which the Convention takes is to list, in Annexes, wastes which cannot be dumped and others for which a special dumping permit is required. The Convention permits dumping to be carried out under certain conditions. The severity of these conditions varies according to the danger to the environment presented by the materials themselves – but there is a 'black list' containing materials which may not be dumped at all. The 1972 Convention is in the process of being replaced by a 1996 Protocol[2], which makes a significant change in the way that it approaches the question of how to regulate the use of the sea as a depositary for waste materials.

The 1996 Protocol, which came into force in 2006 (the 1972 Convention remains in force for those of its parties which have not adopted the 1996 Protocol), is much more restrictive than its parent Convention. Article 4 of the 1996 Protocol states that Contracting Parties 'shall prohibit the dumping of any wastes or other matter with the exception of those listed in Annex 1'. These listed substances are: dredged material; sewage sludge; fish waste, or material resulting from industrial fish processing operations; vessels and platforms or other man-made structures at sea; inert, inorganic geological material; organic material of natural origin; and bulky items primarily comprising iron, steel, concrete and similar non-harmful materials for which the concern is physical impact and limited to those circumstances, where such wastes are generated at locations, such as small islands with isolated communities, having no practicable access to disposal options other than dumping. In terms of 2006 Amendments to the 1996 Protocol, storage of carbon dioxide (CO_2) under the seabed is also allowed.

It is of huge significance to see the way in which the parties to the 1996 Protocol have moved from a list of certain banned substances, with all others being allowed, to a list of certain allowed substances, with all others being banned. This is important recognition of the damage which has been done to the world's oceans

2 1996 Protocol to the Convention on the Prevention of Marine Pollution by Dumping of Wastes and Other Matter, London, 1972, in force 2006.

by most of the world's states treating them as a 'seafill' for centuries, and of the fact that a new approach is needed.

The relevance of this example is to point out that a similar shift is possible within the area of wildlife use and trade. If the IWC could be persuaded, for instance, eventually to list as protected all cetaceans except those to which special considerations apply – in other words, to move from a list of those species that cannot be taken to a list of those that can – environmental protection would be enhanced, and far greater sophistication would result. Likewise, if the parties to CITES could be persuaded eventually to ban *all* commercial trade in wildlife, except for those specific species in respect of which circumstances exist to make their trade environmentally acceptable, then enforcement and implementation would become far easier to control, and environmental protection would significantly be enhanced.

While these suggestions obviously represent a lofty ideal, making far greater use of the CBD (which requires protection of habitats and ecosystems rather than of particular species) would be a way to begin to move toward this scenario. Of course, nothing will happen overnight – as desirable as this would be. We must move in 'baby steps' and a good start would be by way of strengthening the, sadly underutilised, CBD – this would move us closer to the goal of overall protection. What I am arguing for is a subtle, but significant, shift; one which would enable the parties to foster agreement on broad principle, even where they cannot seem to reach agreement on the ground. It should, in addition, be recognised that the CBD is also concerned with improving the welfare of humankind. The best means to strengthen implementation of the CBD would probably be significantly to increase the geographical areas protected on both land and sea – this would mean a move away from the problematic categorisation approach, would be firmly in line with the proposed strengthening of the precautionary principle and would aid inevitably in increasing scientific understanding of biological diversity. Protected areas must become a tool more used – both on land and at sea.

At the same time as strengthening protection of natural biodiversity, we should recognise the value of the number of states which have become involved in the debates over protection and use of whales and elephants. Although the participation of many states without obvious interests has often been criticised as having been for strategic (even dishonest) purposes only, and it can be argued that their presence has caused confusion, it could be an extremely positive step to recognise, in an increasingly globalised world, that all states have a role to play. All states have interests. The principles worked out, over time, in the contexts of whales and elephants, will help to protect also little-known species in little-known countries. Ultimately, this may prove to have been the true value of the leviathan battles fought over the two species.

References

Note: The research in this book has relied heavily on original documents (especially the reports of different CITES CoPs and the Chair's Reports of different IWC meetings) which are available on the official websites hosted by the Secretariat of the Convention on International Trade in Endangered Species of Wild Fauna and Flora (CITES), see www.cites.org; and by the International Whaling Commission (IWC), see www.iwc.int/home. In addition, verbatim recordings and transcripts of IWC meetings are available for recent meetings on compact disc, and for older transcripts in the archives of the IWC in Histon/Impington. Opening statements submitted by contracting governments at recent meetings are available online, and older statements are available in the archives. Such documents are not listed in this list of references.

ABCOnline (2006) 'Anti-whaling Countries Secure Victory', 17 June 2006, www.abc.net.au/am/content/2006/s1665270.htm, accessed 17 June 2006.

Amerasinghe, C.F. (2005) *Principles of the Institutional Law of International Organizations*, Cambridge University Press, Cambridge.

Andresen, S. (2001) 'The Whaling Regime: "Good" Institutions but "Bad" Politics?', in R.L. Friedheim (ed.) *Toward a Sustainable Whaling Regime*, University of Washington Press, Seattle, 235–65.

Anthony, L. (2009) *The Elephant Whisperer*, Sidgwick and Jackson, London.

Bamford, H. (2006) 'Wildlife's Gravy Train', *The Independent*, 21 October 2006.

Berglund, N. (2006) 'Norwegians Cheer Shift on Whaling Ban', 19 June 2006, www.aftenposten.no/english/local/article1357250.ece?service=print, accessed 19 June 2006.

Birnie, P. (1985) *International Regulation of Whaling: From Conservation of Whaling to Conservation of Whales and Regulation of Whale-Watching: Volume I and Volume II*, Oceana Publications, Inc., New York.

Birnie, P. (1985b) 'The Role of Developing Countries in Nudging the International Whaling Commission from Regulating Whaling to Encouraging Nonconsumptive Uses of Whales', *Ecology Law Quarterly*, vol 12, 937–75.

Birnie, P. (2003) 'The Framework for Conservation of Whales and other Cetaceans as Components of Marine Biodiversity', in W.C.G. Burns and A. Gillespie (eds) *The Future of Cetaceans in a Changing World*, Transnational Publishers, Ardsley, 101–39.

Birnie, P. and Boyle, A. (2002) *International Law and the Environment*, Oxford University Press, Oxford.

Black, R. (2005) 'Reform Likely on Whaling Process', 24 June 2005, http://news.bbc.co.uk/go/pr/fr/-/1/hi/sci/tech/4618763.stm, accessed 12 June 2006.

Black, R. (2011) '"Clean-up Bid" Tops Agenda for Whaling Meeting', 10 July 2011, www.bbc.co.uk/news/science-environment-14079749, accessed 25 July 2012.

Blignaut, J., de Wit, M. and Barnes, J. (2008) 'The Economic Value of Elephants', in R.J. Scholes and K.G. Mennell (eds) *Elephant Management: A Scientific Assessment for South Africa*, Wits University Press, Johannesburg, 446–76.

Bonner, R. (1993) *At the Hand of Man: Peril and Hope for Africa's Wildlife*, Simon and Schuster Ltd, London.

Bonner, R. (1994) 'Western Conservation Groups and the Ivory Ban Wagon', in M.M.R. Freeman and U.P. Kreuter (eds) *Elephants and Whales: Resources for Whom?*, Gordon and Breach Science Publishers, Basel, 59–71.

Bowman, M. (1995) 'The Nature, Development and Philosophical Foundations of the Biodiversity Concept in International Law', in C. Redgwell and M. Bowman (eds) *International Law and the Conservation of Biological Diversity*, Kluwer Law International, Deventer, 5–31.

Broch, H.B. (1994) 'North Norwegian Whalers' Conceptualization of Current Whale Management Conflicts', in M.M.R. Freeman and U.P. Kreuter (eds) *Elephants and Whales: Resources for Whom?*, Gordon and Breach Science Publishers, Basel, 203–18.

Burke W.T. (2001) 'A New Whaling Agreement and International Law', in R.L. Friedheim (ed.) *Toward a Sustainable Whaling Regime*, University of Washington Press, Seattle, 51–79.

Burns, W.C.G. and Wandesforde-Smith, G. (2002) 'The International Whaling Commission and the Future of Cetaceans in a Changing World', *Review of European Community and International Environmental Law*, vol 11, no 2, 199–210.

Burton, R. (1980/1973) *The Life and Death of Whales*, Andre Deutsch, London.

caribbeannews.com (2006) 'Science, Not Emotion, Must Influence Whaling Issue Says St Kitts Official', 28 April 2006, www.caribbeannewsnow.com/caribnet/cgi-script/csArticles/articles/000014/001414.htm, accessed 25 May 2006.

Chadwick, D.H. (1992) *The Fate of the Elephant*, Key Porter Books Ltd, Toronto.

Chasek, P.S. (2005) 'Margins of Power: Coalition Building and Coalition Maintenance of the South Pacific Island States and the Alliance of Small Island States', *Review of European Community and International Environmental Law*, vol 14, no 2, 125–37.

Christie, M. (2006) 'Whaling Ban Faces Stiff Test at Caribbean Meeting', 12 June 2006, http://abcnews.go.com/US/wirestory?id=2066893, accessed 15 June 2006.

Clark, C.W. (1983) 'Acoustic Communication and Behavior of the Southern Right Whale (*Eubalaena australis*)', in R. Payne (ed.) *Communication and Behavior of Whales*, Westview Press, Boulder, 163–98.

Clarke, J. (2002) *Coming Back to Earth: South Africa's Changing Environment*, Jacana (Pty) Ltd, Johannesburg.

Clarke, S. (2006) 'The World Today – Pro-whaling Lobby May Have Majority at IWC', *ABC Online*, 16 June 2006, www.abc.net.au/worldtoday/content/2006/s1664679.htm, accessed 16 June 2006.

Clover, C. (2004) *The End of the Line: How Overfishing is Changing the World and What We Eat*, Ebury Press, London.

cnnonlinearchives (2003) 'Whale Gene Research Spawns Big Waves', 25 July 2003, http://cnn.com/2003/TECH/science/07/25/whale.populations.ap/index.html, accessed 5 March 2013.

Couzens, E. (2012) 'Whaling and Dealing: Aboriginal Subsistence Whaling, Politics and

Poverty', in Y. Le Bouthillier *et al.* (eds) *Poverty Alleviation and Environmental Law*, Edward Elgar, Cheltenham, 100–26 .

Creighton, M.S. (1995) *Rites and Passages: The Experience of American Whaling, 1830–1870*, Cambridge University Press, Cambridge.

Dahmani, M. (1987) *The Fisheries Regime of the Exclusive Economic Zone*, Martinus Nijhoff Publishers, Dordrecht.

D'Amato, A. and Chopra, S.K. (1991) 'Whales: Their Emerging Right to Life', *The American Journal of International Law*, vol 85, 21–62.

D'Amato, A. and Hargrove, J.L. (1974) *Environment and the Law of the Sea: A Report of the Working Group on Ocean Environment*, American Society of International Law, Washington DC.

Darby, A. (2007) *Harpoon: Into the Heart of Whaling*, Allen and Unwin, Crows Nest.

Day, D. (1992) *The Whale War*, Grafton, London.

de Klemm, C. and Shine, C. (1993) *Biological Diversity Conservation and the Law: Legal Mechanisms for Conserving Species and Ecosystems*, IUCN Environmental Policy and Law Paper No. 29, Gland and Cambridge.

Douglas-Hamilton, I. (1997) 'Foreword: An East African Perspective', in D. and S. Balfour (eds) *African Elephants*, C. Struik, Cape Town, XX–XX.

Douglas-Hamilton, I. and O. (1992) *Battle for the Elephants*, Doubleday, London.

Eco News (2002), 'CITES: Nations Vote to Protect Minke, Bryde's Whales', 11 November 2002, http://www.unep.org/cpi/briefs/Brief11Nov.doc, accessed 15 July 2013.

Ellis, R. (2003) *The Empty Ocean*, Island Press/Shearwater Books, Washington DC.

ENS News (2003) 'Whaling Commission Authorises Global Conservation', 17 June 2003, www.unep.org/cpi/briefs/BRIEF17June.doc, accessed 15 July 2013.

ENS News (2006a) 'Japan Loses Bid to Resume Commercial Whaling', 18 June 2006, www.ens-newswire.com/ens/jun2006/2006-06-18-01.asp, accessed 18 June 2006.

ENS News (2006b) 'Whale Survival at Stake in War Over Commercial Whaling', 12 July 2006, www.ens-newswire.com/ens/jun2006/2006-06-12-01.asp, accessed 15 June 2006.

European Parliament (2007) 'European Parliament Resolution of 24 May 2007 on the EU Strategic Objectives for the 14th Meeting of CITES', http//eur-lex.europa.eu/Lex UriServ/LexUriServ.do?uri=OJ:C:2008:102E:0451:0455:EN:PDF; 'Summary Note: EP Members in the EC Delegation to CITES COP 14', http://www.europarl.europa.eu/comparl/envi/pdf/delegations/2004-2009/cites-2007.pdf, accessed 30 May 2008.

Fischer, S. (2012) 'Elephant "Speaks" Like a Human', *National Geographic*, 2 November 2012, http://news.nationalgeographic.com/news/2012/11/121102-korean-speaking-elephant-talk-human-science-weird-animals/, accessed 25 January 2013.

Freeman, M.M.R. (2001) 'Is Money the Root of the Problem? Cultural Conflicts in the IWC', in R.L. Friedheim (ed.) *Toward a Sustainable Whaling Regime*, University of Washington Press, Seattle, 123–46.

Freeman, M.M.R. and Kreuter, U.P. (eds) (1994) *Elephants and Whales: Resources for Whom?*, Gordon and Breach Science Publishers, Basel.

Friedheim, R.L. (2001a) 'Fixing the Whaling Regime: A Proposal', in R.L. Friedheim (ed.) *Toward a Sustainable Whaling Regime*, University of Washington Press, Seattle, 311–35.

Friedheim, R.L. (2001b) 'The IWC as a Contested Regime', in R.L. Friedheim (ed.) *Toward a Sustainable Whaling Regime*, University of Washington Press, Seattle, 3–48.

Friedheim, R.L. (2001c) 'Negotiating in the IWC Environment', in R.L. Friedheim (ed.) *Toward a Sustainable Whaling Regime*, University of Washington Press, Seattle, 200–34.

Frost, S. (1978) *Report of the Independent Inquiry into Whales and Whaling conducted by the Hon. Sir Sydney Frost: Volume 1 and Volume 2*, Australian Government Publishing Service, Canberra.

Fry, I. (2005) 'Small Island Developing States: Becalmed in a Sea of Soft Law', *Review of European Community and International Environmental Law*, vol 14, no 2, 89–99.

Gales, N.J., Kasuya, T., Clapham, P.J. and Brownell, R.L. Jr (2005) 'Japan's Whaling Plan Under Scrutiny', *Nature*, vol 435, 883–4.

Gambell, R. (2003) 'I Am Here, Where Should I Be?' in W.C.G. Burns and A. Gillespie (eds) *The Future of Cetaceans in a Changing World*, Transnational Publishers, Ardsley, 65–85.

Gavron, J. (1993) *The Last Elephant: An African Quest*, Flamingo, London.

Gillespie, A. (2002) 'Forum Shopping in International Environmental Law: The IWC, CITES, and the Management of Cetaceans', *Ocean Development & International Law*, vol 33, 17–56.

Glavovic, P.D. (1995) *Wilderness and the Law*, PD Glavovic, Wandsbeck.

Goodman, D. (1997) 'Book Review: P.J Stoett *The International Politics of Whaling*', *Marine Policy*, vol 21, no 6, 547–9.

Goodman, D. (2010) 'Japan's Research Whaling is not Unlawful and does not Violate CITES Trade Rules', *Journal of International Wildlife Law and Policy*, vol 13, no 2, 176–82.

Greenpeace (1997) 'The Pro-trade Syndicate at CITES: Its Secret Names and Tactics', 9–29 June 1997, http://archive.greenpeace.org/comms/97/bio/citesbrief.html, accessed 29 July 2006; http://www.skeptictank.org/treasure/GP1/CITESBRI.TXT, accessed 15 July 2013.

Guinee, L.N., Chu, K. and Dorsey, E.M. (1983) 'Changes Over Time in the Songs of Known Individual Humpback Whales (*Megaptera novaeangliae*)', in R. Payne (ed.) *Communication and Behavior of Whales*, Westview Press, Boulder, 59–80.

Hanks, J. (1997) 'Foreword: A South African Opinion' in D. and S. Balfour (eds) *African Elephants: A Celebration for Majesty*, C. Struik Publishers, Cape Town, XX–XX.

Hardin, G. (1968) 'The Tragedy of the Commons', *Science*, vol 162, 1243–8.

Harland, D. (1994) *Killing Game: International Law and the African Elephant*, Praeger, London.

Hegel, G.W.F. (1821) (Knox, T.M. (trans., 1952)) *Hegel's Philosophy of Right*, Oxford University Press, London.

Heller, P. (2006) 'The Whale Warriors', *National Geographic*, May 2006, www.sea shepherd.org, accessed 26 July 2006.

Hepworth, R. (1997) 'The Independent Review of CITES', paper, University of Nottingham Conference.

HNA (High North Alliance) (1997) 'CITES and the IWC: The Fall of Conservation? The Downlisting of Minke Whales in the North Atlantic', www.highnorth.no/Library/Trade/CITES/ci-an-iw.htm, accessed 12 May 2006; http://www.google.co.za/books/about/CITES_and_the_IWC.html?id=SKx8twAACAAJ&redir_esc=y, accessed 15 July 2013.

HNA (High North Alliance) (2000) 'Trade in Minke Whales – Conservation in Action: Downlisting North Atlantic Minke Whales at CITES, 10–20 April, 2000', High North Alliance, Reine i Lofoten.

Honkonen, T. and Couzens, E. (eds) (2013) *International Environmental Law-making and Diplomacy Review 2011: Synergies Among the Biodiversity-related Conventions*, University of Eastern Finland/United Nations Environment Programme, Joensuu.

Hoyt, E. (2005) *Marine Protected Areas for Whales, Dolphins and Porpoises: A World Handbook for Cetacean Habitat Conservation*, Earthscan, London.

Iino, Y. and Goodman, D. (2003) 'Japan's Position in the International Whaling Commission', in W.C.G. Burns and A. Gillespie (eds) *The Future of Cetaceans in a Changing World*, Transnational Publishers, Ardsley, 3–32.

IISD (2002) 'CITES COP-12 Highlights', 13 November 2002, www.iisd.ca/download/asc/enb2127e.txt, accessed 29 July 2006.

IISD (2006) '54th Meeting of the CITES Standing Committee', 9 October 2006, www.iisd.ca/cites/sc54/, accessed 10 October 2006.

IISD (2007) 'Summary of the Fourteenth COP to CITES: 3--15 June 2007' 18 June 2007, www.iisd.ca/cites/cop14/, accessed 18 June 2007.

The Independent (2006) 'The Great Betrayal: Pro-hunting Japanese Seize Control of Whaling Commission', 17 April 2006, www.independent.co.uk/environment/the-great-betrayal-prohunting-japanese-seize-control-of-whaling-com, accessed 29 May 2006.

Ishii, A. and Okubo, A. (2007) 'An Alternative Explanation of Japan's Whaling Diplomacy in the Post-Moratorium Era', *Journal of International Wildlife Law and Policy*, vol 10, no 1, 55–87.

IUCN Environmental Policy and Law Paper No. 56 (2006) *An Introduction to the African Convention on the Conservation of Nature and Natural Resources*, (2nd ed.) IUCN, Bonn/Gland.

Jackson, P. (1990) *Endangered Species: Elephants*, The Apple Press, London.

Jacobson, J.L. (2001) 'Whales, the IWC, and the Rule of Law' in R.L. Friedheim (ed.) *Toward a Sustainable Whaling Regime*, University of Washington Press, Seattle, 80–104.

Jamaica Observer (2006) 'Japan, Pro-whaling Nations Lose Early Vote at St Kitts Conference', 17 June 2006, www.jamaicaobserver.com/news//107085_Japan—pro-whaling-nations-lose-early-vote-at-St-Kitts-conference.html/, accessed 17 June 2006.

Joy, J. (2001) 'Conservation or the Cannon-Shot Rule', June 2001, www.wob.nf.ca/News/2001/conservation.htm, accessed 30 October 2002.

JWA News (2002) 'Pro-whaling Voices Gain Firmer Ground', June 2002, www.whaling.jp/english/news/0207_01.html, accessed 8 April 2005.

JWA News (2003) 'Japan Should Remain in the IWC and Continue Its Efforts to Reform It', December 2003, www.whaling.jp/english/news/0312_03.html, accessed 8 April 2005.

JWA News (2005a) 'Pro-use Votes Close to Simple Majority at the 57th IWC Meeting in Korea', July 2002, www.whaling.jp/english/news/jwa_news_12.pdf, accessed 7 June 2006.

JWA News (2005b) 'Support Votes Increase for Proposed Alleviation on Minke Whale Trade at CITES CoP 13', January 2005, www.whaling.jp/english/news/jwa_news_11.pdf, accessed 7 June 2006.

Kalland, A. (1994) 'Whose Whale is That? Diverting the Commodity Path', in M.M.R. Freeman and U.P. Kreuter (eds) *Elephants and Whales: Resources for Whom?*, Gordon and Breach Science Publishers, Basel, 159–86.

Komatsu, M. and Misaki, S. (2001) *The Truth Behind the Whaling Dispute*, Institute of Cetacean Research/Japan Whaling Association, Tokyo.

Komatsu, M. and Misaki, S. (2003) *Whales and the Japanese: How We Have Come to Live in Harmony with the Bounty of the Sea*, Institute of Cetacean Research, Tokyo.

Kreuter, U.P. and Simmons, R.T. (1994) 'Economics, Politics and Controversy Over African Elephant Conservation' in M.M.R. Freeman and U.P. Kreuter (eds) *Elephants and Whales: Resources for Whom?*, Gordon and Breach Science Publishers, Basel, 39–58.

Lapointe, E. (2003) *Embracing the Earth's Wild Resources: A Global Conservation Vision* Éditions du Scribe, Sherbrooke, Quebec.

Leakey, R. and Morell, V. (2001) *Wildlife Wars: My Battle to Save Kenya's Elephants*, Macmillan, London.

Leite, J.R.M., Ferreira, H.S. and de Araújo Ayala, P. (2007) 'Protection of Natural Spaces in Brazilian Environmental Law', in N.J. Chalifour, P. Kameri-Mbote, L. Heng Lye and J.R. Nolon (eds) *Land Use Law for Sustainable Development* Cambridge University Press, New York, 471–83.

Lomborg, B. (1998) *The Skeptical Environmentalist: Measuring the Real State of the World*, Cambridge University Press, Cambridge.

Lyster, S. (1985) *International Wildlife Law*, Grotius Publications, Cambridge.

Mackenzie, J.M. (1988) *The Empire of Nature*, Manchester University Press, Manchester.

MacPherson, M. (2001) *The Cowboy and His Elephant: The Story of a Remarkable Friendship*, Headline Book Publishing, London.

Marra, J. (2005) 'When Will We Tame the Oceans?', *Nature*, vol 436, 175–6.

Martin, S. (2001) *The Whales' Journey: A Year in the Life of a Humpback Whale, and a Century in the History of Whaling*, Allen and Unwin, Crows Nest.

McCurry, J. (2006) 'Japan Welcomes Vote Against Whaling Ban', *The Guardian*, www.guardian.co.uk/print/0,,329508679-105193,00.html, accessed 28 September 2006.

McGuinness, P.P. (2002) 'Whale-Huggers' Case is Up the Spout', *Sydney Morning Herald*, 28 May 2002.

McNeill, D. (2006) 'Japan Seizes Control of Whaling Group after Historic Vote', *The Independent*, http://news.independent.co.uk/environment/article1090433.ece, accessed 19 June 2006.

Melville, H. (1851/1994) *Moby Dick*, Penguin Books, London.

Meredith, M. (2001) *Africa's Elephant: A Biography*, Hodder and Stoughton, London.

M'Gonigle, R.M. (1980) 'The "Economizing" of Ecology: Why Big, Rare Whales Still Die', *Ecology Law Quarterly*, vol 9, no 1, 119–238.

Miyaoka, I. (2004) *Legitimacy in International Society: Japan's Reaction to Global Wildlife Preservation*, Palgrave MacMillan, Basingstoke.

Mkoka, C. (2003) 'Wildlife Officials Brace for 2004 Ivory Sale', 12 November 2003, www.mg.co.za/article/2003-11-12-wildlife-officials-brace-for-2004-ivory-sale, accessed 24 March 2004.

MSNBC.com (2006) 'Japan Loses First Round in Whaling Talks', http://msnbc.com/id/13366802/print/1/displaymode/1098/, accessed 17 June 2006.

Mulvaney, K. and McKay, B. (2003) 'Small Cetaceans: Status, Threats, and Management', in W.C.G. Burns and A. Gillespie (eds) *The Future of Cetaceans in a Changing World*, Transnational Publishers, Ardsley, 189–216.

News24.com (2005) 'Botswana Wants Elephant Cull', www.news24.com/News24/Africa/News/0,,2-11-1447_1674548,00.html, accessed 21 March 2005.

News25.com (2004) 'Plan to End Illicit Ivory Sales', www.news24.com/News24/Technology/news/0,6119,2-13-1443_1603429,00.html, accessed 21 March 2005.

Nikonorov, I.V. (1996) *Whale Mystery: A Review*, VNIRO Publishing, Moscow.

Nollkaemper, A. (1996) 'The Legality of Moral Crusades Disguised in Trade Laws: An Analysis of the EC "Ban" on Furs From Animals Taken by Leghold Traps', *Journal of Environmental Law*, vol 18, no 2, 237–56.

Ong, D.M. (1998) 'The Convention on International Trade in Endangered Species (CITES, 1973): Implications of Recent Developments in International and EC Environmental Law', *Journal of Environmental Law*, vol 10, no 2, 291–314.

Osterwoldt, R.U. (1982) 'International Law and Politics of Conservation: The Case of the Whales', M. Phil. (Politics) thesis, University of Oxford.

Parker, I. (2004) *What I Tell You Three Times is True: Conservation, Ivory, History and Politics*, Librario Publishing Ltd, Kinloss.

Pauly, D. and MacLean, J. (2003) *In a Perfect Ocean: The State of Fisheries in the North Atlantic Ocean*, Island Press, Washington DC.

Payne, K. (1999/1998) *Silent Thunder: In the Presence of Elephants*, Penguin Books, New York.

Payne, K. (2003) 'Sources of Social Complexity in the Three Elephant Species', in F.B.M. de Waal and P.L. Tyack (eds) *Animal Social Complexity: Intelligence, Culture and Individualized Societies*, Harvard University Press, Cambridge Massachusetts, 57–85.

Payne, K., Tyack, P. and Payne, R. (1983) 'Progressive Changes in the Songs of Humpback Whales (*Megaptera novaeangliae*): A Detailed Analysis of Two Seasons in Hawaii', in R. Payne (ed.) *Communication and Behavior of Whales*, Westview Press, Boulder, 9–57.

Peta, B. (2004) 'Zim Inflated Its Elephant Population', *Daily News*, 16 September 2004.

Pickover, M. (2005) *Animal Rights in South Africa*, Double Storey Books, Cape Town.

Pohl, O. (2003) 'About-face by Whaling Commission', *The Age*, 18 June 2003, www.theage.com.au/articles/2003/06/17/1055828326653.html, accessed 20 June 2003.

Poole, J. (1996) *Coming of Age with Elephants*, Hodder and Stoughton, London.

Project Jonah (2008) 'An Overview of Solitary Dolphins in New Zealand', www.projectjonah.org.nz/site/projectjon/files/Solitary%20Dolphins%20in%20New%20Zealand.pdf, accessed 20 March 2008.

Raney, A. (2006) 'Majority of Commission Votes to Resume Whaling', *The Washington Post*, 19 June 2006, www.washingtonpost.com/wp-dyn/content/article/2006/06/18/AR2006061800786_pf.html, accessed 19 June 2006.

Reeve, R. (2002) *Policing International Trade in Endangered Species: The CITES Treaty and Compliance*, Earthscan, London.

Ris, M. (1994) 'Conflicting Cultural Values: Whale Tourism in Northern Norway', in M.M.R. Freeman and U.P. Kreuter (eds) *Elephants and Whales: Resources for Whom?*, Gordon and Breach Science Publishers, Basel, 219–39.

Robertson, R.B. (1954/1996) *Of Whales and Men*, Simon and Schuster, New York.

Rose, G. and Crane, S. (1995) 'The Evolution of International Whaling Law', in J. Kirkby, P. O'Keefe and L. Timberlake *The Earthscan Reader in Sustainable Development*, Earthscan, London, 34–9.

Rose, G. and Paleokrassis, G. (1996) 'Compliance with International Environmental Obligations: A Case Study of the International Whaling Commission', in J. Cameron, J. Werksman and P. Roderick (eds) *Improving Compliance with International Environmental Law*, Earthscan, London, 147–75.

Sample, I. (2003) 'Plenty More Fish in the Sea?', *The Guardian*, 15 May 2003, www.guardian.co.uk/science/2003/may/15/science.highereducation, accessed 16 May 2003.

Sand, P.H. (1997) 'Commodity or Taboo? International Regulation of Trade in Endangered Species?', *Green Globe Yearbook 1997*, 19–36.
Sand, P.H. (2008) 'Japan's "Research Whaling" in the Antarctic Southern Ocean and in the North Pacific Ocean in the Face of the Endangered Species Convention (CITES)', *Review of European Community and International Environmental Law*, vol 17, no 1, 56–71.
Sands, P. (2006/2005) *Lawless World: Making and Breaking Global Rules*, Penguin Books, London.
Scales, H. (2012) '"Talking" Whale Could Imitate Human Voice', *National Geographic*, 22 October, 2012, http://news.nationalgeographic.com/news/2012/10/121022-whales-voices-science-animals-humans-marine-mammals/, accessed 25 January 2013.
Scarff, J.E. (1977) 'The International Management of Whales, Dolphins, and Porpoises: An Interdisciplinary Assessment' (*Part One*) and (*Part Two*), *Ecology Law Quarterly*, vol 6, 323–427 and 571–638.
Schücking, H. and Anderson, P. (1991) 'Voices Unheard and Unheeded', in V. Shiva (ed.), World Rainforest Movement *Biodiversity: Social and Ecological Perspectives*, Zed Books Ltd, London, 13–41.
Sea Shepherd (2006) 'Japan Fails to Take Control of the IWC', 16 June 2006, www.seashepherd.org/news/media_060616_2.html, accessed 17 June 2006.
Shaw, M.N. (2003) *International Law*, Cambridge University Press, Cambridge.
Sheldrick, D. (1997) 'Foreword: The Moral Viewpoint', in D. and S. Balfour *African Elephants,* C. Struik, Cape Town, XX–XX.
Sims, D.W. and Southward, A.J. (2006) 'Dwindling Fish Numbers Already of Concern in 1883', *Nature*, vol 439, 660.
Stansell, K.B. (2004) 'Testimony of Kenneth B Stansell, Assistant Director, International Affairs, US Fish and Wildlife Service, Department of the Interior, before the House Resource Subcommittee on Fisheries Conservation, Wildlife and Oceans regarding the international aspects of fish and wildlife conservation', 29 April 2004, www.iipdigital.usembassy.gov/st/english/texttrans/2004/04/20040430171557cmretrop0.7508814.html#axzz2Z705pR00, accessed 29 July 2006; http://www.iipdigital.usembassy.gov/st/english/texttrans/2004/04/20040430171557cmretrop0.7508814.html#axzz2Z705pR00, accessed 15 July 2013.
Stoett, P.J. (1997) *The International Politics of Whaling*, University of British Columbia Press, Vancouver.
Stone, C.D. (1996) 'The Convention on Biological Diversity', in C.D. Stone *Should Trees Have Standing? And Other Essays on Law, Morals and the Environment*, Oceana Publications, New York, 119–33.
Stone, C.D. (2001) 'Summing Up: Whaling and Its Critics', in R.L. Friedheim (ed.) *Toward a Sustainable Whaling Regime*, University of Washington Press, Seattle, 269–91.
Sutherland, G. (1993) *The Whaling Years: Peterhead (1788–1893)*, Centre for Scottish Studies, University of Aberdeen, Aberdeen.
Thomson, R. (1986) *On Wildlife "Conservation"*, United Publishers International, New York.
Thornton, A. and Curry, D. (1992/1991) *To Save an Elephant: The Undercover Investigation into the Illegal Ivory Trade*, Bantam Books, London.
Victor, D.G. (2001) 'Whale Sausage: Why the Whaling Regime Does Not Need to be Fixed', in R.L. Friedheim (ed.) *Toward a Sustainable Whaling Regime*, University of Washington Press, Seattle, 292–310.
Vidal, J. and McCurry, J. (2006) 'Japan Sets Course for Return to Commercial Whaling',

30 May 2006, www.guardian.co.uk/fish/story/0,,1785651,00.html, accessed 6 June 2006.

Vogler, J. (2000) *The Global Commons: Environmental and Technological Governance*, John Wiley and Sons Ltd, Chichester.

Watson, L. (2003) *Elephantoms: Tracking the Elephant*, Penguin Books, London.

Watson, P. (1994) *Ocean Warrior: My Battle to End the Illegal Slaughter on the High Seas*, Key Porter Books, Toronto.

Whitehead, H. (2003) 'Society and Culture in the Deep and Open Ocean: The Sperm Whale and Other Cetaceans', in F.B.M. de Waal and P.L. Tyack (eds) *Animal Social Complexity: Intelligence, Culture and Individualized Societies*, Harvard University Press, Cambridge Massachusetts, 444–64.

Wijnstekers, W. (2007) 'Opening Address: CITES CoP 14', 4 June 2007, IISD, *Earth Negotiations Bulletin* 21: 51.

Yasunaga, H. (2005) 'Pre-modern Whaling in Yobuko and Ogawashima Island', in Institute of Cetacean Research and Japan Whaling Association (eds) *The 3rd Summit of Japanese Traditional Whaling Communities: Report and Proceedings 30 May 2004*, Muroto City and the Institute of Cetacean Research, Muroto and Tokyo, 69–85.

Yeater, M. (2013) 'CITES Secretariat: Synergies Based on Species-level Conservation with Trade Implications', in T. Honkonen and E. Couzens (eds) *International Environmental Law-making and Diplomacy Review 2011: Synergies Among the Biodiversity-related Conventions*, University of Eastern Finland/United Nations Environment Programme, Joensuu, 135–53.

Interviews

Note: Although I have benefited from talking to many people over the years, the following people kindly gave me their time for formal interviews. Their capacities are given as at the date of interview.

Bendikson, Bjorn Hugo (Head, Norwegian Whalers' Union): 26 April 2007; Reine.

Best, Peter (former member of the IWC Scientific Committee): 5 November 2006; Cape Town.

Bloch, Dorete (Director, Natural History Museum, Faroes): 30 April 2007; Tórshavn.

Butterworth, Douglas (member of the IWC Scientific Committee): 17 December 2006; Cape Town (and follow-up emails, 2008).

Donovan, Gregory (Head of Science, IWC): 1 February 2007; Histon/Impington.

Frøvik, Rune (Secretary, High-North Alliance): 26 April 2007, Reine.

Goodman, Dan (delegate to the IWC, Japan): 7 July 2011; St Helier.

Grandy, Nicola (Secretary, IWC): 1 February 2007; Histon/Impington.

Jaakola, Esko (Commissioner to the IWC, Finland): 19 April 2007, Helsinki.

Johannsen, Halvard (Deputy-Director, Ministry of Fisheries, Norway; IWC delegate): 23 April 2007; Oslo.

Kaniaru, Donald (Senior Legal Adviser to UNEP): 29 August 2004; Joensuu.

Kleinschmidt, Horst (former Commissioner to the IWC, South Africa): 7 November 2006; Cape Town (and follow-up emails, 2008).

Mbengashe, Maria (Director, Biodiversity, Dept of Environmental Affairs and Tourism, South Africa): 29 June 2006, Pietermaritzburg.

Mikkelsen, Bjarni (Marine mammal researcher, Natural History Museum, Faroes): 30 April, 2007; Tórshavn.

Morishima, Akio (Chair, Environmental Council for the government of Japan): 7 October 2004; Nairobi.

Morishita, Joji (Alternate Commissioner, spokesman, for Japan to the IWC): 7 July 2011; St Helier.

Nagai, Masa (Senior Legal Officer, UNEP): 4 July 2006; Pietermaritzburg.

Noble, Rachel (Campaigner; Elephants Project, EIA): 6 February 2007; London.

Olsen, Justinus (veterinary surgeon, for the Faroes government): 1 May 2007; Tórshavn.

Oosthuizen, Herman (Commissioner for South Africa to the IWC): 14 December 2006 Cape Town.

Perry, Clare (Senior Campaigner; Whales Project, EIA): 6 February 2007; London.

Rodrigues Eusebio, Turid (Deputy-Director, Ministry of Foreign Affairs, Norway; IWC delegate): 23 April 2007; Oslo.

Rice, Mary (Campaigner; Elephants Project, EIA): 6 February 2007; London.

Rice, Nan (Whale and Dolphin Action Group): 7 November 2006; Cape Town.

Sheldrick, Daphne (Trustee, David Sheldrick Conservation Trust): 2 October 2004; Nairobi.

Sand, Peter (former Secretary-General, CITES): 1 November 2012; Munich (and prior emails, 2008).

Sjúurðarburg, Ólavur (Head, Faroes Pilot Whalers' Association): 29 April 2007; Leirvik.

van Schalkwyk, Marthinus (Minister, Environmental Affairs and Tourism, South Africa): 30 May 2007; Anchorage.

Walløe, Lars (Delegate for Norway to the IWC): 24 April 2007; Oslo.

Watson, Paul (Director, Sea Shepherd Conservation Society): 28 May 2007; Anchorage.

Western, David (former Director, Kenyan Wildlife Service): 8 October 2004; Nairobi.

Note: I also benefited from informal research communications with the following people:

Simon **Brockington** (Secretary, IWC); **Chobe Enclave Trust Community** (Chobe, Botswana); Michael **Cowling** (Dean, Faculty of Law, University of KwaZulu-Natal); Jeremy **Firestone** (Professor, Marine Policy, University of Delaware); Karsten **Klepsvik** (Commissioner to the IWC, Norway); Jennifer **Lonsdale** (Director, EIA); Les **Manley** (Acting Chief Director, Dept of Foreign Affairs, South Africa); Ewan **McIvor** (Senior Environmental Officer, Antarctic Division, Australia); Benson ole **Mutunkei** and Ogeli ole **Makui** (representatives, Kitengela Community Conservation Project, Kenya); Rosemary **Rayfuse** (Professor, Faculty of Law, University of New South Wales); Jeremy **Ridl** (specialist environmental law attorney, South Africa); Bart **Smithers** (environmental documentary film-maker, South Africa).

Index